Concrete Floors, Finishes and External Paving

Philip H. Perkins

CEng, FICE, FCIArb, FASCE, FIWEM

BUTTERWORTH
HEINEMANN

Butterworth-Heinemann Ltd
Linacre House, Jordan Hill, Oxford OX2 8DP

℞ A member of the Reed Elsevier plc group

OXFORD LONDON BOSTON
MUNICH NEW DELHI SINGAPORE SYDNEY
TOKYO TORONTO WELLINGTON

First published 1993
Paperback edition 1995

British Library Cataloguing in Publication Data
Perkins, Philip H.
    Concrete Floors, Finishes and External
    Paving
    I. Title
    693

ISBN 0 7506 2330 6

Set by Hope Services (Abingdon) Ltd
Printed and bound in Great Britain

Contents

Preface

The author wishes to express his thanks to the many people and orga-nizations who have given him help and advice over the years. In particular, his colleagues in the Cement and Concrete Association (now the Concrete Society Advisory Service), the British Cement Association, Keith Green of Burks Green and Partners, consulting civil and structural engineers, Tim Lees now with W.R. Grace Ltd, Roy Harrison of British Ceramic Research Ltd, Jim Tiramani of the National Federation of Terrazzo, Marble and Mosaic Specialists, and Bev Brown of RMC Ltd. However, the opinions expressed in this book are the author's and are based on his experience.

Introduction

It is axiomatic that every building has walls, a floor and a roof. It is recognized that care and attention to specification of workmanship and materials are required for the finishing and surface appearance of the walls, but this recognition is often lacking when it comes to floors and external paving.

This book deals with floors, roofs that are used for car parking, and external paving. One of the difficulties in writing such a book is the extremely wide range of use to which floors are put from the aspects of loading, wear, and resistance to aggressive chemicals. This includes domestic use, offices, department stores, warehouses, garages and workshops, and factories. In addition, there is now a staggeringly large number of proprietary materials on the market for finishing floors for these different uses. The basic structural material for the floor slab is concrete, plain and reinforced, used on its own or in combination with structural steelwork. In this book, the term 'reinforced' includes steel mesh and bars, and fibres (steel and polypropylene). Structural design is not dealt with in this book as there are adequate publications on the subject, but some comments are given on the principles of the design of ground-supported concrete slabs.

Frequent reference is made to British Standards and Codes of Practice as compliance with these documents is desirable. However, it must be remembered that Standards and Codes are compromise documents, particularly now that they have to be in general agreement with Eurocodes, and they set out minimum recommendations. The year of publication of a Standard is not generally given as reference should be made to the latest edition including all amendments.

There is, unfortunately, a decided tendency for Standards, and particularly Codes, to become too long and complex and this creates great difficulty in their interpretation. The author has felt for many years that Codes should provide only the basic skeleton to which the users of the Codes should fill in the necessary details. The present position is that there is less and less scope and encouragement for designers/ specifiers to use their professional judgement, unless they are prepared to risk accusations of professional negligence for failure to comply with the Code should some defects appear in the building. The amount of litigation arising from the design and construction of buildings and structures has increased dramatically during the past 20 years and this is reflected in the very high premiums charged by insurance companies for professional indemnity policies.

One unresolved problem is the extent to which contractors can be held responsible for some design aspects of floor construction and finishes when the specification has been prepared by an independent professional. Also, when the contract specification is very brief, how does this affect the legal responsibilities of the specifier and the contractor? The author assumes that reliance has then to be placed on the general requirement of 'fitness for the purpose'.

There are two basic types of specification: the traditional specification which details materials and workmanship, and the performance specification which details how the finished product must behave under clearly-defined operating conditions and leaves how this is to be achieved to the contractor. While the performance specification has many advantages, it normally leaves open the problem of long-term durability, and does not appear to solve the problem of aesthetic appearance. For how long should the contractor be held responsible for the satisfactory performance of the floor? There is also the related and basic factor of reasonable maintenance. The latter is particularly important because all floors require maintenance, even though many building owners and floor users are unwilling to accept this.

The author is in favour of a practical compromise between a traditional and a performance specification. The contractor must be informed clearly of the materials he has to use and basic requirements for the end product, such as the standard of finish of a floor surface. Any unusual requirements, such as abrasion-resistance testing of the finished surface, and impact-resistance testing on screeds and toppings, must be included in the contract documents and be specifically referred to.

The author believes in competition, but not in the automatic acceptance of the lowest tender, even when tenders have been invited on a selected list. It is a fact that if a contractor finds he is losing money on a contract there is likely to be trouble on site, and in the end one of the losers will surely be the building owner. The author has definite doubts on the inclusion of the mandatory arbitration clause in building and civil engineering contracts.

The author feels it is desirable to draw the attention of readers to the need to observe recognized safety precautions when using certain building materials and all types of equipment. Attention is drawn to the Control of Substances Hazardous to Health Regulations 1988 (COSHH); this requires the manufacturer or supplier to supply information (without being requested to do so), to risks to health or safety arising from the use of substances on site, and customers must be kept up-to-date with revisions of safety data as far as this is 'reasonably practicable'. Advice should be obtained from the Health and Safety Executive as safety requirements are mandatory.

Portland cement when mixed with water is highly alkaline and a 'safety warning' is included as an appendix in all British Standards for cement. Polymer resins are now widely used in construction and are mainly products of the petrochemical industry. There are certain hazards associated with the use of some of these compounds; for details, reference should be made to the manufacturers, and to FeRFA (Federation of Resin Formulators and Applicators, also known as the Trade Federation of Specialist Contractors and Material Suppliers to the Construction Industry), and of course to the Health and Safety Executive. The hazards include: contamination of the skin and eyes; harmful effects of the inhalation of vapour or mist; and fire and explosion. The extent of the hazard varies from one compound to another.

Reference must be made to a British Standard published in 1992, namely, BS 7543: Guide to durability of buildings and building element, products and components. The Standard was many years in preparation and contains much useful information and advice, but the author finds it in some respects rather disappointing. The subject is of course a very subjective one, but more practical even though controversial comments would have been welcome. The author's own thoughts on 'useful life' and 'effective life' are given in the Introduction to Chapter 7.

Chapter 1
Notes on the principal materials used in the construction and finishing of concrete floors and external paving

Introduction

Many of the materials used in the construction and finishing of concrete floors and external paving are well known and are covered by British Standards, and their use/application by Codes of Practice. However, some of the materials used for the surface finish (the flooring), are not covered by Standards and Codes.

Standards lay down basic requirements and tests for specific materials such as cement, aggregates and steel reinforcement. Codes of Practice concentrate on principles of design, construction, workmanship, and acceptance testing.

These notes are intended to give brief information on the basic properties and appropriate uses of the materials referred to. Information on methods of manufacture and the detailed requirements of relevant Standards are not included. When dealing with a material covered by a Standard, reference should always be made to the latest edition of the Standard, including all amendments.

It is now accepted that the term 'cement' can include Portland cement blended with pulverized fuel ash and ground-granulated blast-furnace slag. These pozzolans are not cement in their own right. A specifier should always state what percentage of pozzolan he is prepared to allow as a replacement for Portland cement.

Cements

The principal cements used are: Portland cement (ordinary and rapid hardening) to BS 12; sulphate-resisting Portland cement to BS 4027 and blended cements, namely, Portland pulverized fuel ash cement to

BS 6588, pozzolanic cement with pulverized fuel ash as pozzolana to BS 6610, and ground-granulated blastfurnace slag for use with Portland cement to BS 6699; and Portland limestone cement to BS 7583.

Requirements for the sampling and testing of these cements are given in BS 4550 and EN196: Methods of testing cement. Revised cement Standards were published in November 1991; these revisions follow as far as possible the requirements in the draft of the pre-standard ENV 197. The Standards revised were: BS 12, BS 146, BS 4027, BS 6588, BS 6610 and BS 4246 (which has been redesignated 'High slag blastfurnace cement'). An excellent summary of the major changes in these Standards is contained in an article in the journal *Concrete* (Nov/Dec. 1991) by W. Gutt, Chairman of the BSI Technical Committee on Cement and Lime.

Portland cement – ordinary and rapid-hardening

The chemical composition of Portland cement is complex. The raw materials from which the cement is made consist mainly of lime, silica, alumina and iron oxide. In the kiln, the compounds interact to form very complex compounds, which become even more complex during hydration.

The main compounds of Portland cement are:

- tricalcium silicate   C3S
- dicalcium silicate   C2S
- tricalcium aluminate   C3A
- tetracalcium aluminoferrite   C4AF

For the purpose of chemical analysis to determine the cement content of concrete and mortar, it is usually assumed that the composition of cement ordinary Portland cement (OPC) is 65% calcium oxide and 21% soluble silica. Reference should be made to BS 12 and BS 4550 for requirements on composition, manufacture, and chemical and physical properties. The 1989 edition of the Standard includes requirements for '*controlled fineness*' cement, which has a minimum specific surface of $225m^2/kg$ compared with 275 for OPC and 350 for rapid-hardening Portland cement (RHPC); the compressive strength requirements are somewhat lower. All Portland cements contain gypsum (calcium sulphate), expressed as SO_3; the sulphate content of OPC and SRPC must not exceed 3.5% but for controlled fineness cement the maximum is 3%. The presence of sulphate in these cements is important when interpreting chemical analysis reports on concrete and mortar. The chloride ion content must not exceed 0.1%.

Sulphate-resisting Portland cement

Sulphate-resisting Portland cement (SRPC) is similar in its strength and other physical properties to OPC. The tricalcium aluminate content must not exceed 3%, and the sulphate content must not exceed 2.5%. It is the C3A in the cement that is attacked by sulphates in solution with the formation of ettringite (provided the environment in which the reaction takes place is alkaline). This chemical reaction is expansive in character and it is this expansion which has a disruptive effect on the concrete. Some SRPCs have a lower heat of hydration than some OPCs and are sometimes specified for this reason, but this should only be done after checking with the manufacturers.

Calcium chloride should not be used with SRPC, and in fact it is advisable to check with the manufacturers before using any admixture with this cement.

SRPC is rather darker in colour than OPC and RHPC and the darker shades of SRPC may approach the lighter shades of high-alumina cement (HAC). It is important to bear this in mind when making a visual examination of concrete when the type of cement used is not known.

Blended Portland cements

The relevant British Standards are:

BS 6588: Portland pulverized fuel ash cement
BS 6610: Pozzolanic cement with pulverized fuel ash as Pozzolana
BS 6699: Ground-granulated Blastfurnace Slag (GGBS) for use with Portland cement
BS 7583: Portland limestone cement

Reference should be made to the relevant Standards for complete information on these cements. For sampling and testing, reference should be made to BS 4550; however, at the time of writing, this Standard does not include tests for cement made with GGBS and Portland cement.

Blended cements are used for their low heat characteristic and for their improved sulphate resistance compared with OPC. It should be noted that the rate of gain of strength is significantly lower than that of OPC, but the requirement for initial and final setting times are the same.

In these Standards, the PFA and GGBS can be mixed with the Portland cement on site, and factory blending is not mandatory.

Portland limestone cement is used in Southern Germany and France, where problems are encountered with a lack of fines in the fine aggregate.

Non-Portland type cements

The only non-Portland type cements that may possibly be used for concrete floor slabs and external hard standings are high-alumina cement (HAC) to BS 915.

High-alumina cement

HAC differs from Portland type cements in that it consists predominantly of calcium aluminates.

In the UK, HAC is not permitted to be used for structural concrete. However, HAC is very useful for effecting emergency repairs. Also, it possesses improved sulphate resistance and general chemical resistance compared with OPC. It is darker in colour than OPC and RHPC, but the lighter shades of HAC may be similar to the darker shades of SRPC. The rate of gain of strength is very rapid; the 24-h compressive strength approaches that of a similar mix using OPC. The rate of evolution of heat of hydration is very high and this can cause considerable problems. The water/cement ratio must not exceed 0.40 and to obtain maximum benefit from the use of the cement (the price is about double that of OPC), a high cement content is recommended (400 kg/m^3).

The reason for the prohibition of use for structural concrete in the UK is the phenomena known as 'conversion' which can reduce considerably the compressive strength of the concrete over an appreciable period of time. Before specifying this cement it is recommended that advice be obtained from the manufacturers (Lafarge Aluminous Cement Co., Grays)

Chemically-resistant cements

Chemically-resistant cements are not used in concrete but in mortar for the bedding and jointing of tiles and bricks in situations where a high degree of chemical resistance is required. There are two basic classes of cement, namely resin cement and silicate cement. The resin type cements are mainly based on modified phenolic resins, blended epoxies, furane resin and polyester resin. The silicate cements are based on sodium silicate and potassium silicate and are resistant to high temperatures as well as to a wide range of aggressive chemicals.

For further information on these special cements, reference should be made to the manufacturers, Prodorite Ltd, Wednesbury.

Aggregates for concrete and mortar

There are two basic types of aggregate for concrete floors:

(a) aggregates from natural sources, e.g. quarried and crushed rock, crushed pit and river gravels, and sea-dredged shingle;

(b) artificial lightweight aggregates such as expanded and sintered clays and shales, sintered PFA, foamed blastfurnace slag, and pelletized blastfurnace slag.

Aggregates have a significant effect on a number of the characteristics of concrete, including workability, and this is discussed later in this chapter under the heading 'Concrete as a material'.

Aggregates from natural sources

The relevant British Standard is BS 882, and these aggregates are used on a vastly greater scale than the artificial lightweight aggregates.

The grading limits given in Table 5 for fine aggregate are a reasonable guide, and should be interpreted with common sense. Each sample sieved will yield slightly different results, and allowance should be made for testing error. The sampling and testing of the aggregates are covered by BS 812. The points that require attention are the actual grading of the samples of aggregate being tested; the grading curves should be reasonably smooth (gap-graded aggregate can be accepted in certain circumstances, but generally should be avoided, particularly in fine aggregate used for mortars and screeds). An excess of fines is also undesirable as this tends to increase the water demand of the mix. The width of the grading envelopes for the three grades, C, M and F, have been criticized by some engineers. The author feels that an acceptable grading curve should not run down close to either side of the BS 882 envelope.

BS 882 contains recommendations for limits on the flakiness index, shell content, 10% fines value, and chloride content. Good general advice is given in BS 882: Appendix B: Special considerations.

The 1973 edition of the Standard (Section 4.1: General requirements) is rather more specific in pointing out the types of deleterious matter that can be found in aggregates, e.g. coal and pyrites which can adversely affect the durability of an industrial floor slab. Unfortunately these comments were not carried forward into the revision of 1983.

Artificial lightweight aggregates

Artificial lightweight aggregates are covered by BS 3797, and sampling and testing by BS 3681. Foamed/expanded blastfurnace slag is covered by BS 877.

The main types of lightweight aggregate used for floors are:

Expanded clay	(tradename Leca)
Sintered PFA	(tradename Lytag)
Sintered shale	(tradename Aglite)

Pellite – pelletized blastfurnace slag   (tradename Pellite)
Foamed slag

The principal reasons for using this type of aggregate are to reduce
dead load and provide improved thermal insulation. The abrasion
resistance of concrete made with these aggregates is appreciably lower
than that of concrete made with aggregates to BS 882. Complete technical details of the above aggregates should be obtained from the suppliers.

Steel reinforcement

Steel reinforcement for concrete is covered by the following British
Standards:

BS 4449: Carbon steel bars
BS 4482: Cold-reduced steel wire
BS 4483: Steel fabric
BS 7295: Fusion-bonded epoxy-coated carbon steel bars
BS 4486: Hot rolled, and hot rolled and processed high-tensile and
alloy steel bars for prestressing
BS 5896: High-tensile steel wire and strand for prestressing
BS 4466: Scheduling, dimensioning and cutting of steel reinforce-
ment

The coefficient of thermal expansion of plain carbon steel is 12×10^{-6}.

Epoxy-coated reinforcement

Epoxy-coated reinforcement for which there is now a British Standard
(BS 7295: Parts 1 and 2: 1990) was introduced in the US and Canada
in the mid 1970s. The principal US Standard is ASTM No. A775-81:
Epoxy-coated reinforcing bars.

The epoxy powder is electrostatically applied and is far more resistant to damage than epoxy resin applied as a coating. In the UK there
has been considerable resistance to the use of protective coatings to
steel reinforcement on the grounds that adequate cover of good-quality
concrete provides long-term protection to the rebars. Technically, this
is correct but the large number of defects in reinforced concrete structures of all types, arising from reinforcement corrosion, has amply
demonstrated that in the real world too much reliance should not be
placed on workmanship and site supervision.

Galvanized reinforcement

There is no British Standard for galvanized reinforcement of concrete, but there is a Standard for galvanizing steel, namely, BS 729: Hot-dip galvanized coatings for iron and steel articles. Galvanizing consists of coating the steel with zinc, either by dipping the article in molten zinc or by electrodeposition from an aqueous solution.

The thickness of the coating has an essential influence on durability. Basic recommendations on galvanizing can be obtained from the Zinc Development Association in London and from the International Lead Zinc Organization in New York. There is a chemical reaction between the caustic alkalis in the cement paste and the zinc with the evolution of hydrogen. Generally this has no ill effects but the slight loss of bond which may occur can be virtually eliminated by the application to the galvanized coating of a chromate wash as this will inhibit the chemical reaction between the cement paste and the zinc.

Stainless steel

There are three main groups of stainless steel, namely, martensitic, ferritic and austenitic steels. The austenitic steels are those most widely used in building and construction. Austenitic steel is an alloy of iron, chromium and nickel, and some austenitic steels contain a small percentage of molydenum. The '316' steel contains 18% chromium, 10% nickel and 3% molybdenum and has the highest resistance to pitting and crevice corrosion.

There is a British Standard, BS 6744: Austenitic stainless steel bars for the reinforcement of concrete. The Standard places the bars in two grades, 250 and 460, and two types of steel are specified, type 304 for general applications and type 316 where, as stated above, the highest resistance to pitting and crevice corrosion is required.

Bimetallic corrosion

While bimetallic corrosion is less likely to arise in the construction of reinforced concrete floors than in other parts of reinforced concrete structures, the author feels that some brief comments would be useful, because of the extensive use of stainless steel and non-ferrous fixings.

When dissimilar metals are in direct contact with each other, one of the metals can corrode at the expense of the other. The conditions under which such corrosion can take place are explained in some detail in a BSI publication, PD 6484: Commentary on corrosion at bimetallic contacts and its alleviation. Very briefly, for corrosion to occur, the two metals must be in contact in an electrolyte, e.g. a saline solution such as moisture in concrete.

Current will then flow through the electrolyte from the anodic or baser metal to the cathodic or nobler metal. The anode tends to corrode.

The principle underlying the prevention of corrosion from dissimilar metals in contact is to prevent the flow of current between the two metals. This can be achieved by separating the two metals in such a way that the current does not flow. In reinforced concrete this can be done by ensuring that there is at least 50 mm of concrete between the anode and the cathode, or by providing a suitable coating on both metals where they are in contact or thought likely to be in contact.

The BSI publication (PD 6484) includes tables which indicate the degree of additional corrosion that is likely to occur at the bimetallic contacts listed. To give one example, carbon steels are likely to suffer fairly severe corrosion when in contact with stainless steels in moist concrete. It is prudent to assume that a certain amount of residual moisture will be present in structural concrete.

Admixtures

Introduction

An admixture can be defined as a chemical compound that is added to concrete, mortar or grout when it is being batched or mixed, in order to change some property of the mix and/or the mature concrete, mortar or grout.

The main types of admixtures in general use are:

- accelerators
- retarders
- workability aids
- air-entraining agents
- superplasticizers

Some admixtures perform more than one function; for example, air-entraining agents also act as workability aids. In addition, the following materials can also be considered as admixtures:

- Pigments (BS 1014)
- Pulverized fuel ash (PFA) (BS 3892: Parts 1 and 2)
- Ground-granulated blastfurnace slag (GGBS) (BS 6699)
- Condensed silica fume (at present there is no British Standard for this material)

It is important to remember that the use of an admixture will not turn poor-quality concrete into good-quality concrete.

Accelerators

Accelerators are covered by BS 5075: Part 1. There are in fact two main classes of accelerators: those which increase very considerably the rate of setting of the cement paste, and those which increase the rate of hardening. The former are used in mortar to seal leaks, while the latter are mainly used to offset the effect of low temperatures on the rate of hardening of concrete.

The use of the 'set accelerators' will result in a significant reduction in the strength of the hardened cement paste. The basic accelerator of the second type, for which an entirely satisfactory substitute has yet to be found, is calcium chloride. This compound is no longer used in reinforced concrete due to the danger of corrosion of the rebars, and so chloride-free accelerators appear from time to time on the market, usually with the warning that they should not be used with other admixtures without the prior agreement of the supplier.

Retarders

Retarders are known a set-retarding admixtures as they decrease the initial rate of reaction between the cement and mixing water. For concrete, this type of admixture is covered by BS 5075: Part 1, and for mortar by BS 4887: Part 2.

It should be noted that both Standards require that detailed information on the product shall be provided by the supplier. Retarders are very useful in hot ambient conditions, also when delay is anticipated between mixing and placing, and in the supply of ready-mixed mortars.

Workability aids

This class of admixture is probably the one most widely used in the UK. There are two main types: the lignosulphonates (also known as lignins) and the hydroxylated carboxylic acid salts. They are marketed under proprietary names, each supplier extolling the merits of their own product.

For use in concrete, this class of admixture is covered by BS 5075: Part 1; and for use in building mortars, by BS 4887: Part 1 (it should be noted that this Standard states that the plasticizing admixtures covered are not normally used in screeding).

The great advantage of using a workability aid is that with a constant water/cement ratio improved workability is obtained, or with a constant workability, the water/cement ratio can be reduced. Air-entraining admixtures also act as workability aids and some information on this type of admixture is given in the next section.

Air-entraining admixtures

Air-entraining admixtures for concrete are covered by BS 5075: Part 2, while for mortar the Standard is BS 4887: Part 1. As mentioned previously, an air-entraining agent also acts as a workability aid (plasticizer).

It is now normal good practice to specify air-entrained concrete for all in-situ concrete used for external paving. Very briefly, air-entrained concrete has proved to be more resistant to freeze-thaw and to the effect of the use of de-icing salts. The presence of the entrained air reduces the compressive strength of the concrete and this should be allowed for in the mix design. The actual reduction in compressive strength depends on a number of factors, but an 'average' figure can be taken as 3% for each 1% of air entrained.

Special care is needed if it is proposed to entrain air in cement-rich concrete, i.e. concrete containing more than about 390 kg of cement per m^3. These admixtures are mainly based on vinsol resins or surfactants, and the entrained air is in the form of spherical-shaped voids about 0.02 to 0.25 mm in diameter. A useful side effect is that they also improve the cohesion of the mix, thus making the finishing of concrete slabs somewhat easier.

Superplasticizing admixtures

Superplasticizing admixtures for use in concrete are covered by BS 5075: Part 3. These compounds impart very high workability to a concrete mix, or with a given workability (say 75 mm slump), a very large decrease in water content can be achieved. The increase in workability is dramatic, as the concrete flows so that there is a completely collapsed slump, and the workability is measured by 'flow' as described in BS 1881: Part 105. This 'super-workability' only lasts for a limited period and this is of fundamental importance when using superplasticized concrete. The use of these admixtures has increased considerably in the UK in recent years, mainly in the ready-mixed concrete industry and by firms specializing in the construction of industrial floors of large area. Considerable savings in labour and time in the placing, compacting and finishing the concrete can be achieved.

Pigments

The British Standard for pigments for use with Portland cement and Portland cement products is BS 1014. The Standard lists the types of pigment permitted to be used. In addition to tests on the listed pigments, there are two tests for the effect of the pigment on Portland cement products, namely the effect on setting time and on strength development.

As far as the use of pigmented concrete for floors and external paving is concerned, it is important to remember that complete uniformity and intensity of colour cannot be achieved under normal site conditions. It is therefore recommended that clients be informed of this fact.

Pulverized fuel ash (PFA)

PFA is produced in very large quantities from coal-burning power stations. The relevant British Standard for its use in concrete and grouts is BS 3892: Parts 1 and 2.

It is a very fine powder having a specific surface similar to that of OPC. The main compounds in PFA are oxides of silicon, iron and aluminium, with some carbon, magnesia and sulphur. The Standard limits the sulphur content (expressed as SO_3) to 2.5%, magnesia to 4%, and loss of ignition to 7% when used in structural concrete. Part 2 of the Standard covers PFA for use in concrete for non-structural purposes.

PFA possesses pozzolanic properties and is used as a 'cementitious component' in concrete. This means that it can be used as a cement replacement and thus reduces the heat of hydration and slows down the rate of gain of strength. It imparts some additional sulphate resistance to OPC mixes.

There are two British Standards for the incorporation of PFA with Portland cement: BS 6588: Portland pulverized fuel ash cement, and BS 6610: Pozzolanic cement with pulverized fuel ash as pozzolana.

Ground-granulated blastfurnace slag for use with Portland cement

The relevant British Standard is BS 6699. It covers the use of GGBS for use in combination with Portland cement as a cementitious component of concrete, mortar and grout.

This material possesses pozzolanic properties and also imparts some degree of improved resistance to sulphate attack.

Condensed silica fume

Condensed silica fume is a waste product of the ferrosilicon industry. It consists of 88–98% silicon dioxide, with very small percentages of carbon, ferric oxide, aluminium oxide (alumina), and oxides of magnesium, potassium and sodium. It is an extremely fine, greyish powder, with a specific surface about 50 times that of OPC, and is a highly reactive pozzolan.

The addition of condensed silica fume to concrete and mortar has a significant effect on the properties of the plastic mix and imparts a number of beneficial characteristics to the hardened concrete and mortar. The dosage is likely to be in the range of 2–10% by weight of cement. In the UK and US it is sold in the form of an aqueous suspension. The incorporation of a lignosulphonate plasticizer or superplasticizer is normal practice. Trial mixes are essential, and for important projects, a trial placement under site conditions is very desirable. The 'beneficial characteristics' referred to above include increased cohesion, reduced permeability, increased resistance to sulphate attack and to attack by a number of aggressive chemicals, and an increase in compressive strength.

There is no British Standard for condensed silica fume.

A disadvantage with specifying this material for use in concrete or mortar is that its presence cannot be detected, and so in the event of a dispute there is no clearly defined method of testing for its presence.

Bonding aids

In this book, bonding aids are compounds that, when correctly used, improve the bond between a hardened concrete substrate and cement-based concrete or mortar toppings, and screeds. Thin and thick bed adhesives for use with tiling are dealt with below and referred to in Chapter 4. In the UK the compounds in general use are acrylics, styrene butadiene latex, and polyvinyl acetate (PVA).

In practice, it is particularly difficult to obtain a good, uniform and consistent bond at the interface between the hardened concrete and the fresh concrete or mortar. The compounds listed above are all marketed as proprietary materials and the directions of the suppliers should always be followed.

The following notes are based on the author's experience:

(a) The bond can never be stronger than the weaker of the two materials in contact.
(b) The new material must be laid on the bonding aid while it is still tacky, otherwise it becomes a debonding layer.
(c) It is generally advisable for the base concrete to be well damped down prior to the application of the bonding aid.
(d) Whether or not it is necessary to expose the coarse aggregate of the base concrete by grit blasting or scabbling, is a question on which there are differences of opinion. The author's opinion is that where stress at the interface is likely to be high, it is prudent to provide a mechanical key by lightly exposing the coarse aggregate.

There are two tests for bond, namely the slant shear test which can only be applied in a laboratory, and the pull-off test which can be used on site. Considerable experience is needed in writing a specification for a pull-off test and for assessing the results on site.

Adhesives for tiling

Adhesives for use with tiling are referred to in the relevant British Standards for tiling, namely BS 5385: Parts 3 and 5; and BS 8203: Installation of sheet and tile flooring.

For adhesives used for the tiling covered by BS 5385, reference should be made to BS 5980: Specification for adhesives for use with ceramic tiles and mosaics; the adhesives covered by this Standard (BS 5980) are also suitable for use with terrazzo tiles in those few cases where the terrazzo tiles can be laid with with confidence on a thick-bed adhesive.

The adhesives covered by BS 5980 are cement-based (type 1) and organic-based adhesives (types 2, 3, 4 and 5). They are all proprietary materials and must be used in accordance with the directions of the suppliers.

For sheet and tile flooring (BS 8203), suitable adhesives are covered by BS 5442: Part 1: Adhesives for use with flooring materials. They are all proprietary materials and the choice of adhesive will depend on a number of factors, including the flooring material, type of sub-floor and service conditions. The recommendations of the suppliers must be followed.

Joint fillers and sealants

Introduction

The types of material normally used for movement joints in the structural floor slab are different to those used when these joints are carried through the floor finishes (the flooring). Sealants are of two types:

– in-situ compounds, and
– preformed materials.

Both types should possess the following characteristics:

(a) for external use, and for internal use in floors for wet trades, the sealant should be impermeable;
(b) it should bond well to the sides for the joint;
(c) it must be durable under operating conditions;

(d) as the joint opens and closes, the sealant must respond to that movement without deterioration, fracture or loss of bond to sides of the joint.

Joints in the floor and roof slabs of multi-storey car parks are particularly difficult to make watertight and special proprietary joint assemblies have been developed. At the time of writing this book there were some six British Standards and two Codes of Practice relating to sealants for use in external paving and internal construction. The Standards are listed in the Bibliography at the end of this chapter.

Two particularly useful publications are BS 6213: Guide to the selection of constructional sealants, and BS 6093: Code of practice for design of joints and jointing in building construction. It should be realized that even the best types of sealant are significantly less durable (have a shorter useful life) than the structural material (concrete, brickwork, etc.) in which they are inserted.

In-situ compounds

These in-situ sealants can be separated into four main classes:

- Mastics
- Thermoplastics (hot and cold applied)
- Thermosetting compounds – solvent release
- Thermosetting compounds – chemically curing

The hot and cold applied thermoplastics are mainly used for external paving, and also for floor slabs in industrial buildings.

The thermosetting chemically-curing compounds are mainly the polysulphides, silicones rubbers, polyurethanes and epoxies. While polysulphides are used a great deal, the author's experience is that their performance is unreliable. A relatively new material is epoxy-polysulphide which has a low coefficient of extension, but has good chemical resistance and durability.

Preformed sealants

The majority of the high grade materials are based on Neoprene or Ethylene Propylene Diene Monomer (EPDM). Both are very durable and are particularly resistant to chemical and bacterial attack and are therefore useful for use in special conditions. They are compressed and inserted into the joint with a special tool, after the sides of the joint have been primed and an adhesive applied. The sides of the joint have to be finished very smooth and with considerable accuracy. There is no British Standard for preformed sealants.

Fillers for joints in the base slab

The joint fillers are also known as back-up materials. They are used in expansion joints to provide a base/support for the flexible sealant. The fillers are made from prepared fibres, cellular rubber, and resin-bonded granular cork. Desirable characteristics include durability under operating conditions, and ease in shaping and insertion into the joint. A separating layer of thin sheet material is placed on the top of the filler to prevent the sealant bonding to it.

Sealants and fillers for joints in the finishes

The detailing of joints at the wearing surface of floors requires special consideration. Flexible sealants do not provide support to the arrises of joints. Where the floor is subjected to heavy wear (pedestrian and/or moving loads) the arrises require special protection. There are a number of proprietary mechanical joints on the market, and although they are very costly, they are less costly than having to repair the joints at short intervals. In some cases, a stainless steel cover strip can be used; this should be at least 150 mm wider than the joint and fixed only to one side of the joint.

For contraction and stress-relief joints, plastic or non-ferrous strips can be inserted into the narrow joint, and preferably should be deep enough to extend down to the base concrete. If not, they must be supported on a suitable back-up material.

Ceramic, terrazzo, marble conglomerate (reconstituted marble), marble tiles and composition blocks

Brief information on these materials is given in Chapter 4 as part of the recommendations for laying these materials on concrete and cement/sand screeds. A list of relevant British Standards is given below. It should be noted that there are no British Standards for marble conglomerate and marble tiles, nor for composition block flooring.

- Ceramic floor tiles: BS 6431: Parts 1–6.
- Adhesives for use with ceramic floor tiles: BS 5980
- Terrazzo floor tiles: BS 4131

Adhesives are not normally used for laying terrazzo tiles; before use, advice should obtained from the tile manufacturer or from the National Federation of Terrazzo Marble and Mosaic Specialists.

Magnesite (magnesium oxychloride) flooring

Magnesite flooring is a special type of floor topping consisting of a mixture of calcined magnesite (magnesium oxychloride), selected fine mineral aggregate, fillers, wood flour and pigment, gauged with a solution of magnesium chloride. There is a British Standard, BS 776: Materials for magnesium oxychloride flooring. The Code of Practice for laying this material (CP 204: Section 7) was withdrawn in 1991 as its recommendations were considered out of date and possibly misleading.

Linoleum, PVC, cork and rubber sheet and tiles

Brief information on these materials is given in Chapter 4 as part of the recommendations for the laying of these materials on concrete and cement/sand screeds. A list of the relevant British Standards is given below:

- Linoleum – sheet and tiles: BS 6826
- PVC – unbacked, flexible sheets and tiles: BS 3261
- PVC – backed, flexible sheets: BS 5085
- Cork – carpet sheet and tiles: BS 6826
- Solid rubber flooring – sheet and tiles: BS 1711
- Adhesives: BS 5442: Part 1: Classification of adhesives for use with flooring materials
- General: BS 6263: Code of Practice for care and maintenance of floor surfaces

Agents causing deterioration of concrete materials, excluding corrosion of reinforcement

This section will very briefly discuss acid attack on concrete, alkali-silica reaction, sulphate attack on concrete, and the adverse effects of pyrite in concrete aggregates.

Acid attack on concrete

The Portland cement matrix in concrete is highly alkaline with a pH of about 12.5 to 13.0 and therefore there is a strong chemical reaction when solutions of acids come into contact with it. A solution is acidic when the pH is below the neutral point of 7.0 and alkaline when the pH is above 7.0. The degree of attack depends on many factors, the principal ones being the type of acid, its concentration and pH and the permeability of the concrete. Generally, organic acids tend to be less aggressive than mineral acids. Concrete floor surfaces which may be subject to acidic spillage should be protected by an acid-resistant, durable and abrasion-resistant coating. Figures 1.1 and 1.2 show acid attack on concrete.

Fig. 1.1 Concrete severely etched by an acid spillage.

Fig. 1.2 Deep penetration of concrete by mineral acid.

Fig. 1.3 Concrete affected by alkali-silica reaction. Courtesy: British Cement Association.

Alkali-silica reaction

An alkali-silica reaction is a specific type of alkali-aggregate reaction and was first identified in the USA in about 1940. Since then it has been found to occur in many other parts of the world, including Iceland, Denmark, Germany, United Kingdom, Cyprus, Turkey, South Africa, Australia and Canada.

Since alkali-aggregate reaction was identified as a cause of concrete deterioration in the UK in 1976, and engineers started to look for it as the possible cause of previously-unidentified cracking, the number of reported cases has increased fairly rapidly (see Fig. 1.3).

There are two forms of alkali-aggregate reaction (AAR), namely, alkali-silica reactions (ASRs) and alkali-carbonate reactions. The former (ASR) is much more common the world over and so far is the only type found in the UK. Both types arise from the interaction of alkalis in the cement and with specific types of aggregate.

In the case of ASR the aggregates are siliceous. The key factors are the alkalinity of the pore fluid in the concrete and certain types of siliceous aggregate. The main source of the alkalis in the concrete is Portland cement, but external sources such as sea water and de-icing salts will increase the alkalinity and may raise it to the level at which ASR can commence.

For the purpose of assessing the possibility of ASR, the total alkali in Portland cement is expressed as the equivalent of sodium oxide. It is, however, the total alkali content of the concrete that is the critical factor. The Building Research Establishment (BRE), in their Digest 330 (dated March 1988), suggested that a reasonably safe upper limit for the acid-soluble equivalent sodium oxide per cubic metre of concrete was 3 kg.

The alkalis in Portland cement vary from one factory to another and in the UK the range can be taken as 0.4% to 1.0%. With cement at the upper limit of 1%, the cement content of the concrete would have to be limited to 300 kg/m^3 to meet the BRE recommendation of a maximum of 3 kg/m^3. However, if the alkali content of the cement was 0.6%, the cement content could be increased to 500 kg/m^3, provided that was no other source of alkalis likely to come into contact with the concrete.

The whole subject is a difficult one and expert advice should be sought when using siliceous aggregates, from either the Building Research Establishment, the Concrete Society Advisory Service or the British Cement Association. A few selected publications on this subject are included in the Bibliography at the end of this chapter.

Sulphate attack on concrete

In theory, any solution of sulphates will attack Portland cement concrete to a greater or lesser degree. The sulphate in solution reacts with the tricalcium aluminate (C3A) in the hydrating cement and, in an acidic environment, the compound ettringite is formed, the reaction being expansive in character. As Portland cement concrete is alkaline, it is only under exceptional conditions, such as acid attack on the concrete, that the environment may not be alkaline and therefore suitable for the formation of ettringite.

The degree of attack depends on a number of factors, the principal ones being:

(a) the percentage of C3A in the cement,
(b) the permeability of the concrete,
(c) the solubility of the sulphate, and
(d) whether or not the cations in the sulphate react with compounds in the cement.

Where sulphate attack is thought to be possible, it is prudent to specify sulphate-resisting Portland cement, or to use a pozzolanic admixture in the concrete, such as PFA, GGBS or condensed silica fume. SRPC contains a maximum of 3% of C3A.

The solubility of sulphates varies greatly; calcium sulphate forms a saturated solution at 1100 ppm, while sulphates of sodium, potassium, magnesium and ammonium are very much more soluble.

The cations in magnesium sulphate and ammonium sulphate react with compounds in the cement and double decomposition occurs, which causes these sulphates to be more aggressive to concrete than the others mentioned.

As far as floors are concerned, there are only two likely sources of sulphate: chemical spillage and, for ground-supported slabs only, sulphate in the sub-soil, ground water, and fill used below the slab.

It should be remembered that sulphate attack takes a long time to develop to the stage when damage becomes visible, despite the fact that chemical reaction may commence within a short time of laying a concrete slab on sulphate-contaminated fill, or in contact with sulphates in the sub-soil and/or ground water.

Figure 1.4 shows an example of a severe sulphate attack on concrete.

Fig. 1.4 Surface of concrete attacked by sulphate solution.

Pyrite in concreting aggregates

Until quite recently in the UK it was considered that the presence of iron pyrites in aggregates used for concrete were only important when appearance was a major factor. However, three well-publicized cases have rather changed the situation: one was serious deterioration of the valve tower and link bridge at the Wimbleball dam in Devon; another was the precast concrete parapets on the Bray Viaduct on the A361, North Devon Link Road; and the third was a number of houses in Devon and Cornwall that were constructed of locally-made concrete blocks. In all these cases the aggregate used in the concrete was found to contain appreciable quantities of iron pyrites. The deterioration of the concrete consisted of severe spalling. The aggregates for the valve tower and link bridge and the parapets on the viaduct were from 'natural sources', but the concrete blocks for the houses contained aggregate consisting of material from mining, mainly tin mining.

It is relevant to note that the presence of the chemical compound of iron pyrites does not necessarily mean that damage to concrete made with the aggregate will occur. The actual 'reactivity of pyrite depends upon its chemical structure and form' (Simms, I. and Miglio B. (1990) Mystic Mundic, *New Civil Engineer*, 2 August). Other important factors are the presence of moisture and the availability of oxygen.

The latest (1983) edition of BS 882: Aggregates from natural sources for concrete, contains a warning about the presence of iron pyrites 'where appearance is an essential feature of the concrete'.

Repair can be very difficult, depending on the amount and distribution of the pyrite and environmental conditions. Where the pyrite is contained in discrete pieces of aggregate, it may be possible to cut these out and make good in the usual way. Where the pyrite is widely distributed a durable and acceptable repair may be impossible; replacement of the contaminated concrete may be the only practical solution. The question of legal responsibility for remedial work required by the presence of the pyrite is likewise a difficult one for the Courts.

Concrete as a material

Some rather brief information has been given on the solid constituents of concrete, namely the cement and aggregates; the water used for mixing the concrete has not been commented on as any water suitable for drinking is also suitable for making concrete. In some parts of the world, the water required for making concrete can be a serious problem as the natural sources are contaminated with salt, mainly sulphate and chloride, and are therefore unfit for use. For normal structural concrete, the total acceptable acid soluble SO_3, is usually taken as 4.0% by weight of the cement. As OPC and RHPC are permitted to contain up to 3.5% by mass of sulphur (as SO_3), there is very little margin for additional sulphate in the mixing water and the aggregate. The sulphate limit in OPC and RHPC is reduced to 2.5% if the C3A content of the cement is less than 3.5% by mass. These figures are taken from BS 12: 1989. BS 5328: Part 1: 1990 limits the chloride ion content of reinforced concrete to 0.4% by mass of the cement. In prestressed concrete the limit is 0.10%.

Mix design

The object of mix design is to select and proportion the cement, coarse and fine aggregate, and water in the most economical way so as to achieve the required quality of concrete. The following are the most important of the many factors that influence the quality of the hardened concrete:

(a) the cement content (which determines the aggregate/cement ratio),
(b) the water/cement ratio,
(c) the type, grading, maximum size, shape, and surface texture of the coarse aggregate,
(d) the type and grading of the fine aggregate.

These four factors are inter-related, but even when they have been correctly determined, high quality concrete will only result if the ingredients are properly mixed, followed by correct placing, thorough compaction and curing. The four factors have a fundamental influence on strength and workability.

It is not always appreciated that the grading of the fine aggregate (material passing the 5 mm sieve) has a considerable influence on the properties of the concrete and mortar. In concrete it is generally advisable to reduce the percentage of fine aggregate in the mix when using material of finer grading. The grading envelopes obtained from the grading limits in BS 882 are wide, and the author considers that the mean grading obtained by sieve analysis should lie approximately in the middle of the selected envelope. The use of an aggregate containing an excessive amount of fine material will increase the water demand of the mix.

The workability of the mix determines the ease or difficulty of obtaining maximum compaction. The workability is usually measured by slump and the author is in favour of using a rather higher slump than is often specified in the UK. There is a practical tolerance of ± 25 mm on a specified slump. Higher slumps can be readily obtained with relatively low water/cement ratios by the addition of a plasticizer. Unfortunately, the UK lags behind both the Continent, Scandinavia, and the USA in the use of plasticizers in concrete and mortar.

The need for thorough compaction cannot be over-emphasized, as under-compacted concrete has low strength, high permeability, and low resistance to abrasion.

It is stated elsewhere in this book that abrasion (wear) resistance is directly related to the compressive strength of the concrete. However, where the concrete (base slab or topping) forms the wearing surface, it is sometimes important that the surface should not 'polish' as this can significantly reduce slip and skid resistance. Limestone aggregate is prone to this and should be avoided in concrete where slip resistance is important.

A selected number of references is given in the Bibliography.

Bibliography

Note: where British and other Standards are referenced in detail in the text, they are not included here.

Cements

Concrete Society (1987) *Changes in Cement Properties and Their Effect on Concrete*, Technical Report No. 29

Gutt, W. (1991) British Standards for cements 1991. *Concrete*, Nov./Dec., 31–32

Jackson, P.J. (1989) A decade of cement. *Chemistry & Industry*, 6 Nov., 707–713

Plimmer, J. (1978) *The deterioration of bagged Portland cement in storage*, Advisory Data Sheet 35, Cement and Concrete Association, Feb., p. 2

Prodorite Ltd. (1984) *Corrosion Resistant Cements*, Prodorite Ltd, Wednesbury, p. 31

Aggregates

Building Research Establishment (1991) *Shrinkage of Natural Aggregates*, Digest 357, Jan., p. 4

Collins, R.J. (1986/1989) *Porous Aggregates in Concrete*, Building Research Establishment IP 2/86, p. 4 and IP 16/89, p. 4

Collins, R.J. (1991) *Magnesian Limestone Aggregate in Concrete*, Building Research Establishment IP 2/91, Feb., p. 4

Gutt, W. and Collins, R.J. (1987) *Sea-Dredged Aggregates in Concrete*. Building Research Establishment IP 7/87, July, p. 4

Hammersley, G.P. (1989) The use of petrography in the evaluation of aggregate. *Concrete*, **23** (10), 29–32

Midgley, H.G. (1958) The staining of concrete by pyrite. *Mag. of Concrete Research*, Aug. 1958, 75–78

New Civil Engineer (1990) 'Mundic' attacks Wimbleball dam, 5 July 1990

New Civil Engineer (1990) Mystic 'Mundic' (readers letters), 2 Aug. 1990

Nixon, P.J. (1986) Testing the alkali-silica reactivity of UK aggregates. *Chemistry & Industry*, 21 July 1986, 486–489

Sym, R. (1986) Precision and aggregate tests. *Chemistry & Industry*, 21 July 1986, 484–487

Reinforcement

ASTM (1979) Standard A767–79: Zinc-coated (galvanized) bars for concrete reinforcement

Building Research Establishment (1988) *Corrosion Protected and Corrosion Resistant Reinforcement in Concrete.* IP 4/88, Nov. 1988, p. 4

Cement and Concrete Association (1978) *Galvanized Reinforcement,* Advisory Data Sheet 11, April 1978, p. 8

Clarke, J. (1989) Introducing corrosion resistant reinforcement. *Concrete Quarterly,* Winter 1989, 8–9

International Lead Zinc Research Organization Inc. (1981) *Galvanized Reinforcement for Concrete II,* May 1981, ILZRO, New York, p. 208

Morgan, E. (1990) Quality in steel reinforcement. *Concrete,* **24** (2) 2 Feb. 1990

Roberts, R. (1991) Spacers and cover. *Concrete,* **25** (2), 9–10

Admixtures

Anon. (1976) Superplasticizing admixtures in concrete. Cement & Concrete Association Report No. 45.030, Dec. 1976, p. 32

Brown, B. and Anderson, B. (1988) High strength air entrained concrete. *Concrete Forum,* July, 19–22

Malhotra, V.M. (1981) Superplasticizers – their effect on fresh and hardened concrete. *Concrete International,* May, 66–81

Malhotra, V.M. and Carette, G.G. (1983) Silica fume concrete – properties, applications and limitations. *Concrete International,* May 1983, 40–46.

Male, P. (1989) Properties of microsilica concrete. *Concrete,* **23** (8), Sept. 1989, 31–34.

Bonding aids and adhesives

BS 5350 Methods of test for adhesives, British Standards Institution, Milton Keynes

BS 6319 Testing of resins and polymer-cement compositions. Part 4: Method for measurement of bond strength (slant shear method), British Standards Institution, Milton Keynes

British Board of Agrément (1982) European Union of Agrément Directions for the assessment of ceramic tile adhesives, Aug., p. 28

Coad, J.R. and Rosaman, D. (1986) Site applied adhesives – failures and how to avoid them. Building Research Establishment Report IP 12/86

Dennis, R. (1985) Latex in the construction industry. *Chemistry & Industry*, 5 Aug. 1985, 505-510

Perkins, P.H. (1984) The use of SBR-cement slurry for bonding coats. *Concrete*, **18** (3), 18-19

Plum, D.R. (1990) The behaviour of polymer materials in concrete repair, and factors influencing selection. *The Structural Engineer*, **68** (17), 337–345

Walters, D.G. (1990) Comparison of latex-modified Portland cement mortars. *ACI Materials Journal*, July/Aug., 371–377

Joint fillers and sealants

American Concrete Institute (1970) Guide to joint sealants for concrete structures. *ACI Journal*, July 1970, Title 67-31, 489–536.

Anon. (1987) *Civil Engineering Sealants in Wet Conditions*, CIRIA, Technical Note 128

Anon. (1990) Sealants review. *Concrete*, Oct. 1990, 29–31

BS 5212 Cold applied joint sealants for concrete pavements, British Standards Institution, Milton Keynes

BS 2499 Hot applied joint sealants for concrete pavements, British Standards Institution, Milton Keynes

BS 3712 Building and constructional sealants, British Standards Institution, Milton Keynes

BS 5215 One-part gun grade, polysulphide based sealants, British Standards Institution, Milton Keynes

BS Two-part polysulphide based sealants, British Standards Institution, Milton Keynes

BS One-part gun grade, silicone based sealants, British Standards Institution, Milton Keynes

Beech, J.C. (1981) The selection and performance of sealants. Building Research Establishment IP 25/81

Beech, J.C. and Aubrey, D.W. (1990) *Joint Sealants and Primers – Further Studies of Performance With Porous Surfaces*. Building Research Establishment IP 4/90

PVC sheet and tiles

Building Research Establishment (1971) *Sheet and Tiles Flooring Made from Thermoplastic Binders*, Digest 33 (revised), p. 4

Alkali-aggregate (silica) reaction

Concrete Society (1987) *Alkali-Silica Reaction – Minimising the Risk of Damage to Concrete; Guidance Notes and Model Specification Clauses*, Technical Report 30, p. 40

Sims, I. (1992) The assessment of concrete for ASR. *Journal Concrete*, March/April, 42–46

Sulphate attack on concrete

Building Research Establishment (1981) *Concrete in Sulphate Bearing Soils and Ground Waters*, Digest 250

Browne, R.D. (1986) Deteriology and life prediction – as applied to reinforced concrete. *Chemistry & Industry*, 837–844

Harrison, W.H. (1987) Durability of concrete in acidic soils and waters. *Concrete*, Feb. 1987, 18–24

Perkins, P.H. (1986) *Repair, Protection and Waterproofing of Concrete Structures*, Elsevier Applied Science Publishers, London, p. 302

Robery, P. (1988) Protection of structural concrete from aggressive soils. *Chemistry & Industry*, 4 July 1988, 421–426

Concrete as a material

BS 5328: Parts 1–4, *Concrete*, British Standards Institution, Milton Keynes

BS; DD-ENV 206: 1992 *Concrete, performance, production, placing and compliance criteria*, British Standards Institution, Milton Keynes

Building Research Establishment (1987) *Concrete: Parts 1 and 2*, Part 1: Materials; Part 2: Specification, design & quality, Digest Nos. 325 and 326

General

Building Research Estab. European legislation & Standardisation; Digest 376–Nov. 1992,

Department of the Environment (1989) *Official Journal of the European Communities*, No. L40.12 of 11.2.89

Department of the Environment (1991) *Construction Products Regulations*, Statutory Instrument No. 1620, HMSO, London

Harrison, T. (1992) Update on European Standards, *Concrete Quarterly*, Winter, pp. 16–17.

Chapter 2
Ground-supported floor slabs (slabs on grade)

Introduction

Ground-supported floor slabs can be of in-situ concrete, or plain, reinforced, post-tensioned or precast concrete units. With in-situ concrete, which, except for very small areas, is normally reinforced, it is assumed that the slab is uniformly supported on the ground. The reinforcement is used to control cracking caused by drying shrinkage and early-age thermal stresses arising from (a) restraint (friction) at the interface of the base slab and the sub-base, and (b) thermal and moisture gradients through the slab.

The basis of the design is thus different to that for suspended slabs. If the floor slab is constructed under cover, thermal stresses are unlikely to assume significant proportions and the onset of cracking will occur later than in external slabs. Theoretically, and ideally, drying shrinkage and thus the tensile stresses arising from shrinkage should not start to develop until after the end of the curing period, but it is unwise to rely on this.

Comments on early-age cracking in external in-situ paving are given later in Chapter 5.

Definitions

While some of the expressions used in this chapter are quite familiar to road engineers, they are not so well known in the building industry and therefore these expressions are defined as follows and are illustrated in Fig. 2.1.

- *Sub-grade*: the material immediately below the sub-base, consisting of the natural ground (sub-soil) or imported, selected and compacted fill.
- *Sub-base*: a layer of compacted material between the sub-grade and the base slab.

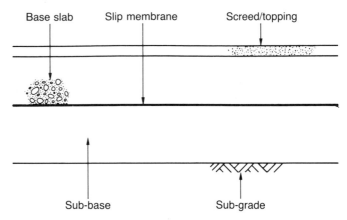

Fig. 2.1 Definitions.

- *Base slab*: also known as the structural floor slab. The top surface of this slab may form the wearing surface of the floor, or the final floor finish may be laid on it. An applied floor finish is known as the 'flooring'.
- *Slip layer*: a layer of material, usually polythene sheeting, laid on top of the sub-base to reduce friction at the interface with the concrete base slab. It also serves to prevent moisture loss from the plastic concrete into the sub-base.

It is clear from the above and from Fig. 2.1 that while the concrete slab is in direct contact with the loads imposed on it, these loads are transmitted downwards through the sub-base to the sub-grade and therefore consideration has to be given to the quality of both the sub-base and sub-grade.

Apart from bearing capacity, the sub-base must not contain any material that is likely to affect adversely the concrete base slab. It is not practical to list all the compounds that may be present in imported fill, but the one most commonly present is sulphate, particularly in material from demolished buildings (e.g. gypsum plaster). Certain shales have also been found to contain sulphate in sufficient concentration to damage and disrupt the concrete floor slabs of houses in the north-east of the UK.

For further information, readers are referred to the Building Research Establishment Digests nos. 363, 274, 275 and 276, which are listed in the Bibliography at the end of this chapter. Reference should also be made to the Building Regulations 1991 and the Approved Documents C1/–/4 made under the Regulations which set out requirements for site preparation and other matters.

Lightly-loaded floors

General comments

Floor slabs of what may be termed 'usual room size' do not need to be structurally designed in the normal sense of the term. They are laid in room sizes of about 15 m^2 to 25 m^2 and are not normally reinforced.

In England and Wales, ground floor slabs are covered by the requirements of the Building Regulations 1991 and the relevant Approved Documents made under the Regulations.

The three basic requirements for this type of floor slab are as follows:

(a) The sub-base and sub-grade must be composed of inert material that will not adversely affect the concrete floor slab and must be capable of safely carrying the loads transmitted from the concrete base slab. Precautions must be taken to avoid danger to health from substances in the ground.

(b) The floor must possess a U-value complying with the requirements of the Regulations.

(c) The floor must comply with the Building Regulations for resistance to the upward passage of moisture and water vapour.

(d) The slab must be constructed of good-quality concrete.

Readers should refer to the Regulations and relevant Approved Documents (ADs) for complete details, but the following observations and recommendations are intended as a guide to good construction.

Sub-soil

See Approved Document C, Table 2 for a list of possible contaminants.

Sub-grade and sub-base

Natural ground (clays) may contain sulphates, and special precautions may be required to prevent sulphate attack on the concrete base slab. Further details on this are given later under 'Concrete for the base slab'. Filled ground is always suspect and should be investigated for the possible presence of aggressive compounds as well as for bearing capacity. In many cases it may be necessary to remove the fill and lay the sub-base on the natural ground.

Suitable material for fill for sub-grade and sub-base is selected hardcore (sulphate free) and granular material. The hardcore should be broken into pieces of 50 mm maximum size and the sub-base should be blinded with sand or other inert fine material. The sub-base must be well compacted. Reference should be made to the Building Regulations 1991 and to Approved Document C1/–/4: Site preparation, sub-soil drainage, contaminants, resistance to weather and ground moisture. The 1985 version has been replaced by the 1992 edition. Table 2 of Section 2 in Approved Document C lists possible contaminants that may be present in the ground, and suggests relevant action to prevent danger to health and safety.

For floors exceeding about 4 m × 4 m it is advisable to lay a slip membrane of 500 gauge polythene sheeting on the blinded sub-base, but *this should not be considered as the damp-proof membrane required by the Building Regulations and the Approved Documents made thereunder.*

Thermal insulation

The relevant Approved Document is 'L', which took effect from 1 April 1990; it deals with the following requirements of Part L, of Schedule 1 to the Building Regulations 1991.

Requirements
Reasonable provision shall be made for the conservation of fuel and power in buildings.

Limits of application
For ground-supported concrete floors the maximum U-value permitted by the AD is 0.45 W/m²K. Heat loss from this type of floor is largely at the perimeter of the floor and therefore special attention should be paid to the insulation of the perimeter. Diagram 2 in Approved Document L shows this very clearly, in as much that floors having dimensions in excess of about 20 m × 20 m do not need insulation in order to meet the maximum U-value of 0.45 W/m²K.

Figure 2.2 shows a suitable arrangement for a domestic floor, which should have a U-value less than 0.45 W/m²K.

Exclusion of moisture from the ground

The exclusion of moisture from the ground is dealt with in Approved Document C which requires that the walls, floors and roof of the building shall adequately resist the passage of moisture to the inside of the building. There is no restriction on the limits of application and so it could be assumed that this requirement applies to all buildings. However, Section 3, clause 3.2 allows a limited number of exclusions which are not particularly clear and encourage different interpretations by different Building Control Officers. It would appear to exclude

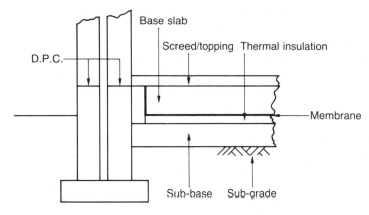

Fig. 2.2(a) Floor with thermal insulation below base slab.

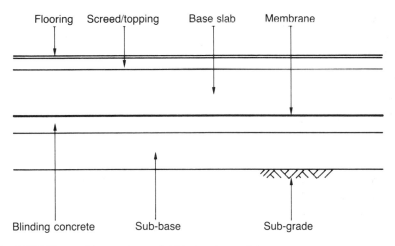

Fig. 2.2(b) Floor with membrane laid below base slab.

warehouses and 'wet' trades. The Document also requires that a floor next to the ground should prevent moisture from the ground reaching the upper surface of the floor. 'Moisture' includes water vapour as well as liquid water.

This requirement is of particular importance as concrete is essentially a 'porous' material, i.e. it possesses a pore structure and allows vapour to pass through it. Materials of this type which are used for floor slabs and toppings, and include magnesite (magnesium oxychloride) toppings, are known in the trade as 'breathing' floors. In effect this means that if moisture slowly penetrates upwards from a damp sub-soil it will evaporate from the top surface of the floor unless the surface is sealed with and bonded to a material that is resistant to the transmission of water vapour; this is dealt with in more detail in Chapters 3 and 4.

A 'technical solution' described in Approved Document C4, is for the provision of a damp-proof membrane located below or above the concrete floor slab. This membrane, if a polyethylene sheet, should be at least 1000 gauge, lapped 150 mm and the joints sealed. Figure 2.2 shows the general construction.

An alternative to sheeting is to use an in-situ coating, preferably in three coats. If brush applied, the second coat should be applied at right angles to the first coat, and applied to the top surface of the concrete floor slab.

All proprietary materials, whether sheet or coating, should be applied in accordance with the manufacturer's recommendations. It is important that the base on which the damp-proof membrane (dpm) is applied should be reasonably smooth, otherwise the sheet material is likely to be damaged, or an excessive thickness of an in-situ coating will be required to cover the high spots. In effect this means that the concrete should be power floated or trowelled, as a tamped finish is likely to be too uneven for a satisfactory dpm to be laid upon it.

This damp-proof membrane must be carried up the perimeter walls to at least the level of the dpm in the walls. This is shown in Figure 2.2.

While most materials used for dpms require protection after laying and are not suitable for use as wearing surfaces, a limited range of proprietary epoxy and polyurethane resins used in the form of a topping about 4-6 mm thick, may be considered as an effective dpm and forms an abrasion-resistant surface.

Concrete for the base slab

Approved Document C4 (1992 edition) makes recommendations for the concrete mix; however, the author suggests that reference is made to BS 5328: Part 1: 1991, Table 13 and to the relevant clauses in Part 2: 1991.

The concrete should be well compacted, and cured for 3 days by covering with well-lapped polythene sheets held down around the perimeter by blocks or scaffold boards.

As mentioned above, the concrete slab should be laid on a slip membrane if the slab size exceeds about 4 m × 4 m. This is to reduce friction at the interface between the concrete and the sub-base; it also helps to prevent loss of moisture from the concrete into a dry sub-base.

Reference has been made to possible sulphate attack on the concrete from sulphates in the sub-grade (sub-soil) and/or sub-base. The recommendations in Building Research Establishment Digest No. 363 should be followed as these set out the mimimum requirements for the concrete.

Heavily-loaded floors for commercial and industrial use

Introduction

The experience of the author is that heavily-loaded floors in commercial and industrial settings give rise to more complaints by users than any other part of the building structure. Any repairs, other than isolated small areas, interfere with the business carried on in the building and can cause inconvenience and financial loss to the occupier. There is no doubt that some complaints are unjustified. Some users refuse to accept that all floors require maintenance and when, as a result of hard use, defects start to appear, they consider they have been let down by their advisers and the contractor, and may resort to court action.

A fundamental point in the design, specification and construction of ground-supported reinforced concrete slabs is that the reinforcement is there to *control* cracking – not to prevent it occurring.

In 1988 the Concrete Society, in association with the British Industrial Truck Association and the Storage Equipment Manufacturers Association, published a comprehensive report on concrete industrial ground floors. It is not the intention of the author to repeat the details given in the Concrete Society publication, but to emphasize certain important aspects of industrial ground floor design and construction, some of which are not detailed in the Concrete Society Report No. 34. The author also makes reference to publications by the Portland Cement Association (PCA) of the USA.

Repairs to concrete floor slabs are dealt with in Chapter 7. See Chapter 5 for comments on cracking in concrete slabs.

Notes on the principles of design and construction of ground-supported slabs to carry heavy moving and point loads

Essentially, there are four ways to design/construct the slab:

(a) the conventional method used largely in the UK, known as the 'long strip' method, using fabric reinforcement;
(b) construction of the floor in relatively large bays with minimum joints, using fabric or bar reinforcement, sometimes with the addition of randomly-dispersed fibres;
(c) construction of the floor in large bays with randomly dispersed steel/polypropylene fibres in the concrete; conventional steel reinforcement is not used;

(d) construction of the floor in relatively large bays using a post-tensioning technique.

The floors being considered here include warehouses and stores where the moving loads are heavily-loaded fork-lift trucks and goods stored directly on the floor or on some form of racking. With racking systems the loads are transmitted to the concrete floor slab as a series of point loads; the concrete floor slab then transmits these loads to the sub-base and sub-grade.

Racking in warehouses

With high racking the resulting point loads are likely to be a major factor in the design of the floor slab. The consequent use of high turret type trucks will require the floor to be laid to very close tolerances of level and flatness.

A special form of high-density storage consists of mobile racking. This is a system of racking mounted on small steel or polyurethane wheels which run on steel rails embedded and anchored into the floor slab; the racking being driven by electric motors. Brief comments on this system are given later in this chapter.

The detailed design of such floors is outside the scope of this book and readers who need this information should refer to the Bibliography at the end of this chapter. However, it is of the greatest importance for the designer to have complete information on the anticipated moving and static loads on the floor. The load-carrying capacity of the concrete will be reduced if the point loads from high racking are near the perimeter of the bays and formed joints. Thus the layout of the racking needs to be known at the design stage for maximum economy in design.

Conventional design

The basic principle of the design of slabs on the ground using conventional reinforcement (fabric or bar) is to design the reinforcement to cater for the 'sub-base friction', but taking into account heavy point loads in the manner recommended in the Concrete Society Report No. 34 and the relevant publications of the PCA in USA (see the Bibliography at end of this chapter). The concrete mix design should be such that the concrete to be placed in position possesses the following basic characteristics:

(a) The degree of abrasion resistance needed for the use to which the floor will be put (a characteristic strength of not less than 40 N/mm^2 for general industrial use and 50 N/m^2 for heavy industrial use), with corresponding cement contents and suitable aggregates. Reference can usefully be made to BS 5328, particularly Part 1: 1991, Table 13; the mix selected should be a designated mix, probably RC40 or RC50, and from this one should go to Part 2: 1991, Section 13 – Designated mix.

(b) Adequate workability for the method of compaction and finishing to be used (see Figs 2.3 and 2.4).

Fig. 2.3 Large industrial ground-floor slab under construction using razor back screeds on permanent concrete supports. Courtesy: Monk Construction Ltd.

Fig. 2.4 Large industrial ground-floor slab; levelling with laser level. Courtesy: Monk Construction Ltd.

Other design principles

If random steel fibres are used without conventional rebars as in the
'Eurosteel/Silifiber' process, the slab is designed against flexure, which
is different to the sub-base friction principle. The promoters of this
system claim that the slab can be thinner than with conventional
design and the spacing of contraction/stress-relief joints can be
increased considerably. The weight of steel fibre in the concrete mix is
likely to be between 20 and 35 kg/m^3.

The bearing capacity of the selected sub-grade and sub-base must
also be taken into account. Some basic information on the selection of
imported fill material has been given under the section in this chapter
dealing with lightly-loaded floors. For an assessment of bearing capacity, reference should be made to appropriate publications listed at the
end of this chapter.

The author is indebted to Eurosteel for the information which
follows on the specification and construction of floors containing
randomly-dispersed steel fibres.

- Characteristic strength: 30 N/mm^2
- The aggregate grading must be carefully controlled and the per-
  centage of sand (fine aggregate passing the 5 mm sieve) is stated
  to be in the range 45–60%
- Slump: 75 mm at the batching plant
- Maximum water/cement ratio: 0.55
- Minimum cement content: not stated, presumably controlled by
  the water/cement ratio, strength and slump

A superplasticizer is added to the mix in the truck mixer as soon as it
arrives on site, with the drum revolving at maximum speed, to increase
the slump to about 175 mm. The use of a special dispenser (a blastmachine pump) is recommended to achieve maximum dispersion of the
superplasticizer (see Figs 2.5 and 2.6).

The special Eurosteel fibres are then dispersed into the mix in the
truck mixer using a Eurosteel blastmachine to prevent 'balling' of the
fibres. The quantity of fibres is likely to be in the range 20–35 kg/m^3 of
concrete. After the dispersion of the fibres, the slump is likely to fall to
about 150 mm, but this will depend on the weight of fibres added to
the mix.

The actual slump required at the time of placing will depend on the
equipment used for levelling and compacting the concrete. A laser level
is normally used for levelling large floor areas (see Fig. 2.4). The minimum slab thickness is 125 mm.

To ensure complete embedment of the fibres in the concrete the surface of the slab should be lightly tamped with a vibrating straight-edge
as soon as possible after placing and levelling.

Fig. 2.5 Custom-built steel fibre dispenser. Courtesy: Monk Construction Ltd.

Fig. 2.6 Silifibres being dispensed into ready-mixed concrete truck. Courtesy: Monk Construction Ltd and Eurosteel s.a.

The slab is then power floated at the appropriate time, which is normally determined by an experienced operator. A high quality curing membrane should be used and applied as soon as the finishing operations have been completed. Rectangular bays should have a length to width ratio not exceeding 1.5. For large floor areas, the bays are usually square with a maximum area of about 2500 m².

Stress relief joints are normally sawn at a carefully determined time after casting, and should be sawn to a depth of ⅓ to ¼ of the depth of the slab, and are usually 3 mm wide. In rectangular bays, the sawn joints are located at about 6 m centres; in square bays they are located at about 6 m centres in both directions Free movement joints should be provided around the perimeter of the floor and around columns and other projections through the floor slab.

In large unobstructed floor areas, there is some lack of guidance on the location of free movement joints and the specifier has the choice of accepting the recommendations of the fibre supplier or using his own technical judgement, based on accepted good practice.

It is unfortunate that there is no recognised design standard for fibre reinforced concrete floor slabs. See also comments in Chapter 5.

Joints

General considerations

It is not possible to construct a very large area of concrete floor without any joints; however, joints can be a potential source of trouble, due to the deterioration (fretting) of the edges of the joints and leakage through the joints when the floor is used for a 'wet' trade.
Unfortunately, a probable alternative to the provision of joints is the formation of cracks. However, very narrow cracks (up to about 0.5 mm wide) are likely to give less trouble than joints.

It is reasonable to reduce joints to a minimum, but basic principles of good slab design must be followed; these principles are set out in detail in the Concrete Society Technical Report No. 34 and also the publications of the Portland Cement Association of the USA and the Cement and Concrete Association of Australia; these are listed in the Bibliography at the end of this chapter. The comments given here apply to conventionally-designed slabs using steel fabric or bar reinforcement.

Types of joints

- *Expansion joints* are intended to isolate structurally the concrete on each side of the joint (Fig. 2.7).
- *Contraction joints* include stress-relief joints, and longitudinal

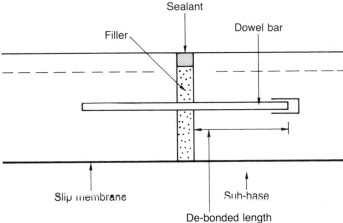

Fig. 2.7 Sketch of an expansion joint.

warping joints (which also act as stress-relief joints). These are intended only to open as the concrete on each side of the joint contracts (Figs 2.8 and 2.9).

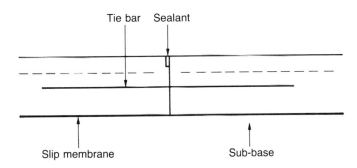

Fig. 2.8 Sketch of a warping joint.

- *Isolation joints* are normally provided at the perimeter and around projections through the floor such as columns and bases for equipment and machinery, manholes, etc. (Fig. 2.10).

Some brief comments on each type of joint follow:

Expansion joints
Expansion joints should be detailed and constructed to allow movement of the concrete in both directions at right angles to the line of the joint, although the greater part of the movement is likely to arise from contraction rather than from expansion. This type of joint is only needed when the floor exceeds about 100 m in length or width,

provided the floor slab is separated from the building structure by an isolation joint. Expansion joints are normally provided with dowel bars for load transfer (see Fig. 2.7).

Further brief comments on dowel bar assembly are given in the next section.

Contraction joints, including stress-relief joints and warping joints

Contraction joints are all detailed to allow the concrete on each side of the joint to contract, so that the joint tends to open fractionally. The reinforcing fabric located about 50 mm from the top surface of the slab is stopped off 50-75 mm each side of the joint. Tie bars are provided, located at the neutral axis, and usually consist of 12 mm diameter bars at about 500 mm centres (Figs 2.8 and 2.9).

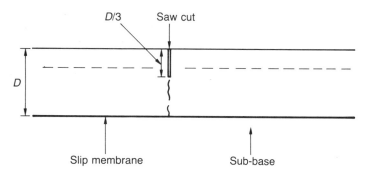

Fig. 2.9 Sketch of a sawn stress-relief joint.

The decision on whether to use tie bars (which are bonded throughout their length) or dowel bars (which are debonded for half their length), is a matter for the designer. The ends of the bays (the length of the 'bay' is determined by the area of longitudinal reinforcement provided) are normally provided with load-transfer dowel bars, debonded on 50% of their length.

For the satisfactory functioning of dowelled joints, it is essential that they are carefully and clearly detailed and that equal care is taken in their assembly on site to ensure secure fixing of the dowel bars in their correct position. If this is not done, cracks are liable to form near the joint and rectification is expensive. Dowel bars must be fixed in the horizontal position and at right angles to the line of the joint.

It is also essential that the debonded sections of the bars are all located on the same side of the joint. If for any reason the joint becomes 'locked', crack formation is almost inevitable. In the case of longitudinal warping joints, when the side forms are removed, it is not unusual to find some honeycombing along the exposed side of the slab. This should be patch repaired or if the depth affected is very shallow, a coat of bituminous emulsion can be applied; the object is to reduce bond between the concrete of the new bay and the concrete of the 'old' bay.

Transverse stress-relief joints are usually formed with saw cuts to a depth of at least one-quarter of the depth of the slab, but the author recommends one-third (0.33%) of the slab depth. To make a clean cut, the concrete must be sufficiently mature, but if the sawing is delayed, a crack may have already formed near the proposed line of the cut (Fig. 2.9).

An alternative is a 'wet-formed' joint, created by inserting a plastic strip in the concrete as part of the finishing operation. A disadvantage is that as the joint widens with time, the insert becomes loose and the arrises tend to break down. The use of a crack inducer in the bottom of the slab, extending upwards for a distance of one-quarter of the slab thickness, has lost favour in the UK in recent years. This is due to problems of workmanship because if the crack inducer is not securely and accurately fixed below the intended saw cut or wet-formed joint, it is likely to result in the formation of a crack in the surface of the slab near the wet-formed or sawn joint. Some recommendations for repairs to joints and cracks are given in Chapter 7.

Isolation joints

It is unfortunate that isolation joints are often omitted from floor design. As a result of this omission it is quite usual to find that cracks have appeared radiating from the corners of protrusions through the slab (e.g. columns, machine bases, manholes, gulleys, etc.) (Fig. 2.10).

It is also advisable for square and rectangular columns to have a plinth at the base with the corners orientated to line up with the bay

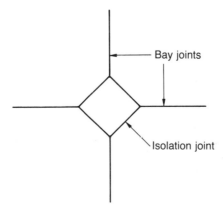

Fig. 2.10 Sketch of an isolation joint around a column/manhole slab.

joints. With machinery bases and manholes it may not be possible to arrange for this line-up and then extra reinforcement should be provided.

Isolation joints are difficult to seal effectively against water penetration so the number should be kept to a minimum in floors used for wet trades. When they are omitted, extra reinforcement should be provided at the corners at the same level as the main fabric.

Bay layout and bay sizes

If the floor is to be used as a warehouse, it is essential for the designer to obtain full and detailed information on the moving loads that will operate on the floor and the layout and loading of the racking. In real life, it is unfortunate that this vital information is not always available at the time of the design of the floor. When this occurs, the designer should point this out in writing to the client and list the possible consequences. The Concrete Society Technical Report No. 34 emphasizes that: 'The capacity of the slab will be reduced when the loads are close to the edge' [of a bay].

It is better to use steel road forms to define the perimeter of bays rather than timber formwork as it is important for the bay edges to be formed as straight as possible. The author is not in favour of proprietary precast concrete forms which are left in position and form part of the floor. By using these forms, two joints are created instead of one (one on each side of the precast concrete form). Also there is no readily available way of checking whether the quality of the concrete in the precast form is as high as that specified and provided in the concrete of the in-situ floor slab.

Gradients, tolerances on surface levels, abrasion resistance, slip resistance, chemical resistance

Floor gradients

Normally, floor slabs are only given falls, or laid to specific gradients if a 'wet' trade is intended in the building. The term 'wet trade' is defined here as a trade where appreciable quantities of water or other liquids will be spilt onto the floor surface, and/or the floor will be frequently washed and the water/liquid has to be disposed of quickly and efficiently. To achieve this desirable state of affairs requires not only effective gradients to the floor surface but also adequate drainage, and a reasonably smooth surface.

The establishment of wet trades usually requires special surface finishes to the structural slab; in-situ finishes are dealt with in Chapter

3. The author recommends that whenever possible, the required gradients (falls to drainage points) should be provided in the structural slab rather than in the screed or topping.

Experience in the construction of concrete roof slabs has shown quite clearly that ponding is inevitable unless the gradients are 'adequate'. For example, if it is decided that a gradient of 1 in 80 is required to provide effective run-off, then an overall design gradient of 1 in 40 should be provided if ponding is to be avoided. However, such a gradient is likely to be too steep for the safety of workers, and then the factory owner would have to accept some degree of ponding. It must be realized that even a few millimetres depth of water on a floor will be seen as a 'pool of water'.

A smooth finish to the floor is desirable for ease of cleaning. There is thus a conflict between the need for slip resistance and ease of cleaning. This is discussed later in this chapter and in Chapters 3 and 4.

It should be noted that 'gradient' is different to 'flatness' and this is discussed in the next section.

Tolerances on surface levels: flatness

Specified tolerances for the levels of the finished surface of the structural slab in cases where this is used as the wearing surface are also known as 'flatness control'. Reference should also be made to the section on tolerances in Chapter 3.

Figure 2.11 illustrates the difference between 'gradient' and 'flatness', which has a direct bearing on 'ponding' (see the previous section) and the very close tolerance required by high racking and the use of narrow-aisle high-lift turret trucks.

This subject is covered in detail in the Concrete Society Technical Report No. 34, which was written in association with the British Industrial Truck Association and the Storage Equipment Manufacturers Association. However, the author feels that some general comments are justified.

There are four principal methods for describing/specifying floor surface tolerances/regularity:

(a) The method described in the Concrete Society Technical Report No. 34.
(b) The US method, known as the F-Number System which is detailed in ASTM E1155.
(c) The German system described in DIN 18 202.
(d) The comparatively simple method described in BS 8204, Appendix A, which utilizes a 3 m long straight-edge and stainless steel 'slip gauges'.

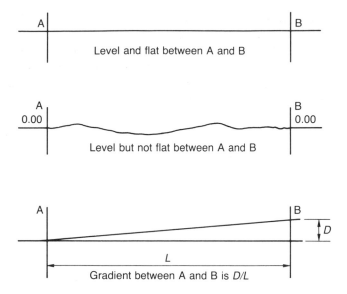

Fig. 2.11 Diagrams illustrating the terms 'level', 'flat' and 'gradient'.

The fundamental point is that it is absolutely essential that tolerances on levels for the control of flatness should be defined in the contract and be relevant to what is actually required for the satisfactory operation of the floor. The specifier must include in the contract full details of how the tolerances he specifies will be checked and what action will be taken if they are not achieved. The specifier must also keep in mind that the closer the tolerances demanded, the higher the cost of constructing the floor; there is no need to waste money on unnecessarily close tolerances. The contractor is entitled to know exactly what is required and how it will be controlled. For further details, see the Bibliography at the end of this chapter.

Abrasion resistance

It is necessary to remember that the abrasion resistance of a concrete floor slab is a function of the top few millimetres of the slab. For slabs that have not received surface grinding, this top layer is composed mainly of cement (the binder), and fine particles of the 'fines' in the concrete.

A direct and practical method for determining the likely abrasion resistance of a concrete floor slab is to carry out a rebound hammer survey; the rebound hammer readings will be a measure/indication of the hardness of the surface. Rebound hammer readings are often used as an indirect measure of the compressive strength of the concrete, but the author's experience is that such results should be regarded as indicative only. About 15 readings per bay would be reasonable. This method is referred to again in Chapter 6.

A more comprehensive method of checking the abrasion resistance of a concrete slab is to carry out a comprehensive survey and testing by means of special equipment developed by the Cement and Concrete Association (now British Cement Association). This is referred to again later in this section.

The factors which significantly influence abrasion resistance are:

(a) binder content and type of binder
(b) water/cement ratio
(c) compaction
(d) finishing
(e) curing
(f) special surface treatment, sealants, hardeners, etc.

Binder content and type. The term 'binder' is used to include both plain Portland cement and blends of Portland cement with PFA and GGBS. Tests carried out by the Cement and Concrete Association and reported on by Chaplin, showed that when using blended cements, particularly those containing GGBS, special care must be taken with curing to help ensure high abrasion resistance; in other words, blended cements containing GGBS are more adversely affected by poor curing than when plain Portland cement is used as the binder.

Water/cement ratio. The reduction in compressive strength resulting from increasing the water/cement (w/c) ratio is well known and is widely documented. Other factors being equal, the same applies to abrasion resistance, i.e. the higher the w/c ratio, the lower the abrasion resistance.

Compaction. Under-compacted or inadequately-compacted concrete will have a relatively low compressive strength and low abrasion resistance.

Finishing. The method of finishing the slab has a direct and profound influence on the abrasion resistance of the floor surface. Finishing with a vibrating tamper will provide an acceptable skid-resistant finish to floors subject to pneumatic tyred vehicles, but is not suitable for the average industrial floor which requires high abrasion resistance. The principal finishing techniques that are employed are:

– hand trowelling, for relatively small areas only, and
– power floating followed by power trowelling.

The two should not be confused. The former is a preliminary finishing process intended to provide a reasonably even surface that is suitable for power trowelling. The timing of carrying out power floating is important and can only be decided by experience; cement content, ambient temperature and relative humidity will influence the time lag between completion of compacting the concrete and commencement of power floating. There is a further period between completing of power floating and commencement of power trowelling, and this again can only be assessed by experience. These 'waiting periods' are required to allow the bleed water to come to the surface and evaporate, and for the surface to stiffen; sometimes two and even three successive trowellings are needed to provide a really dense, hard and abrasion-resistant surface.

To reduce the time interval between completion of compaction, including the initial levelling and striking off of the slab, and the commencement of power floating, vacuum dewatering can be employed. This is of course an additional expense, but when time is the problem, vacuum dewatering can provide an effective answer. The promoters of the system claim that the water content of the concrete is reduced for its full depth, thus increasing the compressive strength of the concrete and the abrasion resistance of the surface layer.

Curing. To produce a floor slab with a high resistance to abrasion, it is essential to pay special attention to curing which is particularly important when using blended cement containing GGBS. The two curing methods that are recommended are polythene sheeting, well lapped and held down around the perimeter and kept in position for at least 4 days, and a high-quality spray-applied resin-based membrane. The latter should not be used if it is intended to apply a surface sealant or other finish. The curing regime should start immediately after the finishing process is complete.

Surface treatments. The final application of surface sealers/hardeners is optional, and in theory are not required. However, the author's experience indicates that the use of a relatively low cost material such as Lithurin (magnesium silico fluoride) can be very useful.

To obtain a high abrasion-resistant surface on a concrete floor slab, the following general recommendations apply:

- a relatively high binder content (about 400 kg/m^3);
- a low 'free' w/c ratio, not exceeding 0.45;
- aggregates to BS 882; flakiness index should not exceed 35, and an aggregate impact value should not exceed 25%. The use of limestone fines should be avoided. The aggregate should be free of contaminants such as lignite and iron pyrites;
- the characteristic strength of cubes made and tested in accordance with BS 1881: Parts 108 and 116, should be 40 N/mm^2. The characteristic strength of cores taken from the floor should not be less than 27 N/mm^2 (BS 1881: Part 120);
- the slump should be 75 mm ± 25 mm;

- the concrete should be thoroughly compacted, normally by a single or double vibrating beam; the use of poker vibrators may be necessary around the perimeter of the bays.

As a more sophisticated method of test for abrasion resistance than the rebound hammer, consideration can be given to the use of the special equipment developed by the Cement and Concrete Association (now the BCA). This is claimed to simulate the type of wear that a floor is likely to experience under the action of heavy trucks, trolleys, etc., running on very hard wheels. It is capable of producing measurable wear in a reasonable period of time (say 15 min) on a high-quality floor slab. This enables an acceptable area of floor to be tested in one day. The test is reproducible. The author would like to see it incorporated in a British Standard or Draft for Development.

Work carried out by Chaplin and published by the BCA indicates that for high abrasion resistance a floor slab without any surface treatment (sealants, etc.) should have a rebound hammer index of not less than about 45, and a wear depth not exceeding about 0.20 mm. The detailed report by Chaplin is included in the Bibliography at the end of this chapter.

Figure 2.12 shows special equipment developed by the British Cement Association (Wexham Developments Ltd, Crowthorne) for assessing abrasion resistance on site.

Slip resistance

Ordinary good-quality concrete possesses good slip resistance which means that the frictional resistance between the surface of the dry concrete and the shoes of persons walking on the floor is sufficiently high to prevent sliding or slipping. It is the characteristics of the contact surfaces that determine the slip resistance. It therefore follows that the characteristics of the soles of the footwear are also important, particularly when the surface of the concrete becomes contaminated by oil, grease or fine powder, or becomes smooth with use.

The resistance to slip is in reality the friction at the interface of the two materials, i.e. the concrete and the soles of the footwear used by the workers.

The changes to the concrete surface which result in changes in the coefficient of friction are:

(a) polishing of the surface under the action of traffic;
(b) contamination of the surface (by oil, grease, wax, fine powder, etc.) arising from the work carried out;
(c) a combination of (a) and (b).

Fig. 2.12 Equipment for measuring abrasion resistance on site. Courtesy: British Cement Association.

The frictional resistance of a rough surface is obviously greater than that offered by a smooth surface. The finishing operations carried out to ensure a high-quality abrasion-resistant surface are also likely to produce a smooth surface that is potentially slippery when wet or contaminated, as in (b) above.

A considerable amount of research has been carried out into the frictional resistance between various flooring materials and various materials used for the soles of industrial footwear. The leading authorities in this field are:

British Ceramic Research Limited (BCRL)
Rubber and Plastics Research Association (RAPRA)
Shoe and Allied Trades Research Association (SATRA)

The BCRL developed a very useful and practical piece of equipment to measure the coefficient of any type of floor surface on site, and this is known as the TORTUS; it can be purchased or hired. Figure 2.13 shows the TORTUS.

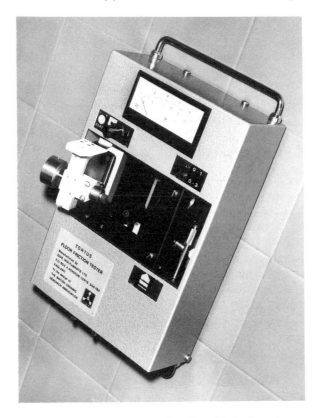

Fig. 2.13 Slip-resistance (measurement of surface friction) equipment. Courtesy: British Ceramic Research Ltd.

The Transport and Road Research Laboratory (TRRL) developed a piece of equipment for measuring skid resistance which can be used on roads and external paving.

There are sometimes conflicting requirements, namely to provide a floor that has a high coefficient of friction to avoid slipping, and to provide a floor surface that is readily and easily kept clean, as in food processing plants. A compromise has then to be found. This should be made quite clear to the building owner at the time the specification is being prepared.

Chemical resistance

The concrete discussed in this book for floor slabs and external paving is based on Portland cement and this type of cement is attacked by acidic solutions having a pH value of 6.0 and below. Other chemicals in solution are also likely to attack Portland cement concrete, and for a detailed list of chemicals in common use the reader is referred to the report by the American Concrete Institute, No. ACI 515.IR79; a detailed reference is given in the Bibliography at the end of this chapter.

While most aggressive chemicals that come into contact with ground-supported concrete slabs are derived from the trade carried on in the building, some attention should be paid to the possibility of chemically-contaminated sub-soil and ground water. Irrespective of their origin, the following chemicals in solution may attack concrete; the degree of attack depends mainly on the concentration, contact time, and temperature:

- acids, particularly mineral acids (sulphuric, hydrochloric, etc.);
- sulphates in solution, particularly ammonium and magnesium;
- fruit and vegetable juices;
- lactic acid derived from milk;
- sugars;
- ammonium compounds except ammonium carbonate;
- distilled and demineralized water;
- compounds containing phenols;
- caustic soda above 10% concentration will slowly disintegrate concrete, mainly by crystallization rather than chemical attack.

While protective coatings are usually necessary, the first line of defence is a low-permeability, high-quality concrete. The author has found that the addition of styrene butadiene rubber (SBR) emulsion to a concrete or mortar mix will reduce to some extent the degree of attack by aggressive chemicals. The extent of the protection given will depend on the chemical involved, its concentration, contact period and temperature. To be effective the SBR should be used in the proportion of 10 litres of emulsion to 50 kg of cement. It also acts as a plasticizer and so reduces the water demand of the mix when the workability (slump) is kept constant.

Special problems arising from the installation of mobile racking

It has been found that the use of mobile racking instead of fixed racking results in an increase of warehouse storage space of up to about 25%.

The pallets are placed in the racking in the usual way by turret trucks; the racking is carried on mobile bases which run on rails cast into the floor. Details of this racking are beyond the scope of this book and the author has not found an independent publication dealing with this type of racking; readers are therefore referred to the designers/manufacturers or to an organization such as the NMHC Consulting Group at the Cranfield Institute of Technology.

It is the fixing of the rails in the floor slab that can cause problems. Two types of rails are used: guide rails and intermediate rails.

The rails can be supported on 'temporary' steel supports and the concrete for the floor slab cast around the rails in the usual way using poker vibrators for the compaction of the concrete under the rail flanges and around the web and bulb. With this method, which is the one usually adopted, there is bound to be a limited amount of entrapped air under the flanges and against the web.

Such floor slabs are likely to be 200–250 mm thick and if the depth of the rails is about 60 mm, this leaves 140–190 mm of concrete below the base of the rails. This would normally contain additional reinforcement to that in the top of the slab between the rails. The presence of this reinforcement increases the difficulty of placing and compacting the concrete around the rails.

The rails must be very accurately and securely fixed and care should be taken to ensure that they are not disturbed during the casting of the concrete slab.

The concrete has to be accurately finished flush with the top surface of the rail flanges, as the turret trucks travel at high speed up and down the narrow aisles between the racking.

An alternative method is to fix the rails in rectangular ducts formed when the floor is cast. The rails can then be accurately supported on steel shims and high-strength fine-aggregate concrete thoroughly compacted under the flanges and against the web and bulb.

It is essential that the rails are checked for alignment and levelness as soon as possible after concreting is complete.

Figure 2.14 shows a mobile racking system in a large warehouse.

Fig. 2.14 A view of mobile racking in a large warehouse. Courtesy: Bar Pro Storage Systems Ltd and NMHC Consulting Group.

Bibliography

Publications and papers detailed in the text are not included here.

ACI 117: 1990 Standard specifications for tolerances for concrete construction and materials, American Concrete Institute, Detroit, USA

ACI 302 IR80: 1980 Guide for concrete floor and slab construction, American Concrete Institute, Detroit, USA

ACI SCM-5(83): 1983 Design of industrial floors, American Concrete Institute, Detroit, USA

ASTM E 1155: 1987 and ASTM M 115: 1987 Test method for determining floor flatness and levelness using the F number system, American Society for Testing Materials, Philadelphia, USA

BS 5328: Parts 1–4: Concrete, British Standards Institution, Milton Keynes

BS 6699 Ground granulated blastfurnace slag for use with Portland cement, British Standards Institution, Milton Keynes

BS 6399: Part 1: Loading for buildings, Code of Practice for dead and imposed loads, British Standards Institution, Milton Keynes

BS 5606 Code of Practice for accuracy in building, British Standards Institution, Milton Keynes

BS 8204: Part 1: Code of Practice for concrete bases and screeds; Part 2: Code of Practice for Concrete wearing surfaces, British Standards Institution, Milton Keynes

BS 6089 Guide to the assessment of concrete strength in existing structures, British Standards Institution, Milton Keynes

BS 8110: Parts 1 and 2: Structural use of concrete: British Standards Institution, Milton Keynes

BS 4483 Steel fabric for reinforcement of concrete, British Standards Institution, Milton Keynes

BS 4449 Carbon steel bars for reinforcement of concrete, British Standards Institution, Milton Keynes

BS 5896 High tensile wire and strand for concrete prestressing, British Standards Institution, Milton Keynes

BS 4486 Hot roller high tensile alloy steel bars for concrete prestressing, British Standards Institution, Milton Keynes

BS 4446 Scheduling, dimensions, bending and cutting steel reinforcement for concrete, British Standards Institution, Milton Keynes

BS 882 Aggregates from natural sources for concrete, British Standards Institution, Milton Keynes

BS 812 Testing aggregates, British Standards Institution, Milton Keynes

BS 6213 Guide to selection of constructional sealants, British Standards Institution, Milton Keynes

BS 2499 Hot applied joint sealants for concrete pavements, British Standards Institution, Milton Keynes

BS 5212 Cold poured joint sealants for concrete pavements, British Standards Institution, Milton Keynes

BS 4254 Two-part polysulphide based sealants, British Standards Institution, Milton Keynes

CP 102 Code of Practice for Protection of buildings against water from the ground, British Standards Institution, Milton Keynes (partially replaced by BS 8102: 1990)

BS 2832 Hot applied damp resisting coatings for solums, British Standards Institution, Milton Keynes

BS 12 Portland cement, ordinary and rapid-hardening, British Standards Institution, Milton Keynes

BS 4027 Sulphate resisting Portland cement, British Standards Institution, Milton Keynes

BS 6588 Portland pulverized fuel ash cement, British Standards Institution, Milton Keynes

BS 6610 Pozzolanic cement with pulverized fuel ash as pozzolana, British Standards Institution, Milton Keynes

BS 4550: Parts 1–6 Methods of testing cement, British Standards Institution, Milton Keynes

BS 3892: Parts 1 and 2 Pulverized fuel ash for use as cementitious component in structural concrete, British Standards Institution, Milton Keynes

BS 1014 Pigments for Portland cement and Portland cement products, British Standards Institution, Milton Keynes

BS 5075 Concrete admixtures, British Standards Institution, Milton Keynes

BS 6044 Specification for road marking paints, British Standards Institution, Milton Keynes

Armitage, J.S. and Judge, C.J. (1987) *Floor Loading in Warehouses – A Review*, Building Research Establishment Report 109, p. 18

Building Research Establishment (1987) *Classes of Imposed Floor Loads for Warehouses*, IP19/87, BRE, p. 4

Building Research Establishment (1981) *Concrete in Sulphate-Bearing Soils and Ground Waters*, Digest 250

Building Research Establishment (1983) *Fill: Parts 1 and 2*, Digests 274 and 275

Building Research Establishment (1983) *Hardcore*, Digest 276

Building Research Establishment (1987) *Concrete: Parts 1 and 2*, Digests 325 and 326

Building Research Establishment (1979) *Estimation of Thermal and Moisture Movements and Stresses: Parts 1, 2 and 3*, Digests 227–229

Building Research Establishment (1990) The U-values of ground floors – application to the Building Regulations IP3/90

Chandler, J.W.E. and Neal, F.R. 91987) *The Design of Ground Supported Concrete Industrial Floor Slabs.* Interim Technical Note 11, Cement & Concrete Association, Wexham Springs

Chapman, R.G. (1986) *The Influence of Cement Replacement Materials and Other Factors on the Abrasion Resistance of Concrete*, Technical Report 568, Cement & Concrete Association, Wexham Springs

Anon. (1988) CCL tackle their largest ever post-tensioned floor. *Concrete*, June 1988, pp. 30-31

Concrete Society (1988) *Concrete Industrial Ground Floors*, Technical Report 34, p. 112

Concrete Society (1987) *Changes in the Properties of Ordinary Portland Cement and Their Effects on Concrete*, Technical Report 29, p. 80

Deacon, R.C. (1987) *Concrete Ground Floors, Their Design, Construction and Finish*, Cement & Concrete Association, Wexham Springs, p. 23

Deacon, R.C. (1991) Welded steel fabric in industrial ground floor construction. *Concrete*, Nov/Dec 1991, 41–44

Department of Environment (1991) Building Regulations 1985 and Approved Documents C & L, HMSO, London, p. 21

Department of Transport (1992) *Specification for Highway Works* 7th edn, HMSO, London

Garber, G. (1991) *Design and Construction of Concrete Floors*, Edward Arnold, London

Gaskill, C.A. and Jacobs, R.C. (1980) *Utilization of Shrinkage-Compensating Concrete – A Cold Storage Warehouse Application*, SP64, 13-31, American Concrete Institute, Detroit, USA

Grohmann, P. (1981) Ribbed slab floor of reinforced concrete in cold-storage depot. *Beton*, **31** (1), 11–12 (in German)

Harrison, R. and Malkin, F. (1983) On-site testing of shoe and floor combinations. *Ergonomics*, **26**(1), 101–108

Malkin, F. and Harrison, R. (1980) A small mobile apparatus for measuring the coefficient of friction of floors. *Journal of Physics D*, 77–79

Spears, R.E. (1983) *Concrete Floors on Ground*, 2nd edn, Portland Cement Association Skokie, Illinois, USA, p. 36

Packard, R.G. (1976) *Slab Thickness Design for Industrial Concrete Floors on Grade*, Portland Cement Association (USA) p. 16

Walker, B. (1990) *Fast-Track Concrete Paving*, Reprint No. 1/90, British Cement Association, Crowthorne, p. 2

Chapter 3

Finishes for concrete floor slabs: directly-finished slabs, screeds, toppings, in-situ terrazzo and over-slabbing floors and roofs for car parking and cold-store floors

Introduction

There is a wide range of methods and materials available for finishing concrete floor slabs. The type of finish selected should depend on the use to which the floor will be put and the amount of money the building owner is prepared to spend. This will determine whether the structural concrete slab is to form the wearing surface or whether an additional layer or layers have to be put down on the structural slab. These finishes can be very expensive indeed. For suspended-floor slabs, the type of finish must be decided and specified at the structural design stage as the finish will affect the dead load of the floor.

During the past 20 years in the UK, considerable attention has been paid to developing methods for speeding up the finishing of concrete slabs so as to obtain a surface finish that will:

(a) provide a satisfactory wearing surface, or
(b) be suitable for the direct laying thereon of thin sheet/tiles without the introduction of a screed.

A fundamental point to be remembered is that all cement-based materials used for finishing floor slabs will be subjected to a certain amount of drying shrinkage which may result in the formation of fine cracks, and will generate a certain amount of dust under wear. While drying shrinkage can be controlled/reduced, it cannot be completely eliminated, and the same applies to 'dusting'. These two characteristics are of great importance in certain industries where neither fine cracks nor any degree of dusting is acceptable.

In selecting a finish, consideration should be given to the structural design of the floor and the supporting structure, in relation to deflexion and general rigidity of the structure under load. The structural behaviour of the building will have a significant effect on the performance/behaviour of the selected finish.

There is an increasing tendency to construct commercial and industrial buildings with steel frames and composite floors or precast concrete floor units with a relatively thin screed or topping. The effect of these forms of construction on rigid types of floor finish is discussed later in this chapter.

It must be kept in mind that the movements which occur in a large building, e.g. shopping centre, multi-storey car park, etc., cannot be predicted accurately because so many factors are involved. With good design and construction, these movements will not adversely affect the structural stability of the building but will affect the finishes to the floors and the potential watertightness of a flat concrete roof slab, of floors in a multi-storey car park, and floors where a 'wet' trade is carried on.

Definitions

For the purpose of this book the following definitions apply (Fig. 3.1):

- *Floor finish*: the finish given to the surface of an in-situ concrete floor slab. This may be the final finish and thus form the wearing surface, or it may be the finish required to make the surface suitable to receive an additional layer or the flooring.
- *Flooring*: the top, final layer of the floor system which is intended to form the wearing surface; also sometimes referred to as floor covering.

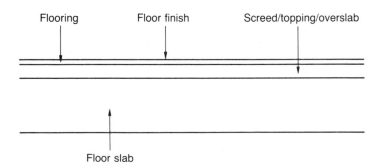

Fig. 3.1 Definitions

- *Screed*: a well-compacted layer of material, usually, but not always, based on Portland cement, applied *in-situ* to a concrete base slab and finished to receive the flooring. A screed as defined in this book is not considered suitable to form the wearing surface of the floor.
- *Topping*: a well-compacted layer of material applied *in-situ* to a concrete base slab. It generally forms the wearing surface, and is well bonded to the base slab.
- *Overslab*: a well-compacted layer of material, normally not less than 100 mm thick, applied *in-situ* to an existing concrete floor slab, either to make up levels or because the existing base slab is not in a fit state to receive a screed or topping. It is usually not bonded to the base slab and is often laid on a separating layer to eliminate reflected cracking and ensure structural separation.

Further definitions can be found in BS 4049: Glossary of terms applicable to internal plastering, internal rendering and floor screeds; and in BS 6100: Glossary of building and civil engineering terms, with special reference to Sub-section 1.3.3.

Direct finishing of concrete slabs

General considerations

Some information will be given on methods of obtaining an abrasion-resistant surface to concrete floor slabs, together with some notes on tolerances.

Tolerances

Reference should also be made to the Tolerance section in Chapter 2.
The recommendations for tolerances of concrete floors for what may be termed general use, are contained in two British Standards, namely BS 5606: 1990: Guide to accuracy in building, and BS 8204: Part 2: In-situ floorings – Code of Practice for concrete wearing surfaces.

Brief information from these two Standards is given below, and was valid at the time this book was written. Such information should always be checked with the latest edition/revision of the Standards. Reference should also be made to the Bibliography at the end of this Chapter and Chapter 2.

BS 5606: 1990, Table 1, recommends:

- Non-suspended floor slabs:
  - thickness: ± 10 mm
  - variation from target plane: ± 25 mm

 – Suspended structural floor:
   variation from target plane: ± 25 mm

There are notes attached to the table, one of which is a definition of a suspended structural floor which is 'one designed to span between edge supports'.

BS 8204: Part 2: 1987, recommends three gradings for maximum permissible departure under a 3 m straight-edge in contact with the floor:

high standard: 3 mm variation
normal standard: 5 mm variation
utility standard: 10 mm variation

As BS 8204 deals exclusively with floors, the author considers that the recommendations in this Standard be followed. In selecting the standard of flatness for inclusion in the contract specification, the specifier should realize that the closer the tolerance, the higher will be the cost of achieving it. Therefore the high-standard tolerance of 3 mm should only be imposed when it is necessary for satisfactory operation of the floor. Even so, for warehouses containing high racking and turret trucks, the specifier should discuss in detail the tolerances required with the supplier of the equipment that will be used on the floor and with the building owner. The selected contractor must be made fully aware of the tolerances required, the reasons for them, and the consequences of failure to meet them.

Once a concrete floor has been constructed, it is a major disaster to find that the surface tolerances are too wide for the safe and efficient operation of the mobile equipment used on it. The experience of the author indicates that there are often misunderstandings over the meaning of the terms, 'level', 'flat' and 'gradient'/'slope'. Figure 2.11 in Chapter 2 illustrates these terms and reference should be made to the section in Chapter 2, Tolerances on surface levels: flatness.

Special requirements for warehouses using turret trucks

It is not unusual for special requirements to be demanded by the suppliers of turret trucks which operate in warehouses with narrow aisles. These special requirements normally relate not only to 'flatness' but also to the need for a reasonably uniform coefficient of friction on the floor surface to avoid wheel spin and skid in the operation of the turret trucks.

A typical requirement tolerance on 'flatness' would be:

 – no sudden irregularities, and

– any gradual irregularities must not contain a slope exceeding 1.5 mm in 300 mm.

A floor achieving the above tolerances would be known as a 'super-flat' floor.

The author feels that the requirements of some UK manufacturers of narrow-aisle turret trucks may be unnecessarily rigid and do not make proper allowance for the difficulty and cost of meeting these tolerances. Experience on the Continent appears to suggest that somewhat wider tolerances could be used without reducing the efficiency and safety of the operation of the turret trucks. To reduce the cost and difficulty of providing very close tolerances on 'flatness', some changes in the design and construction of narrow-aisle turret trucks may be needed, but this is something which does not appear to commend itself to the manufacturers.

Regarding the friction characteristics of the floor, the author has not seen any data for friction coefficients quoted nor demanded, but he understands that the aisles should have the same type of finish throughout. This should result in a reasonable uniformity of grip between the wheels of the turret trucks and floor surface, making acceleration and braking safer.

Methods in general use for direct finishing of concrete slabs

As previously stated, the method adopted will depend on the use to which the floor will be put and whether any additional layers are specified to be applied to the structural slab. This section considers the direct finishing of the structural floor slab.

In this case the surface of the structural slab forms the wearing surface of the floor. The comments which follow apply to slabs supported on the ground (slabs on grade).

The simplest type of finish is by tamping with a vibrating beam tamper working off the side forms. To avoid arguments on site, the Contract specification should state clearly the standard of finish required, i.e. whether the finish direct from the tamper is acceptable or whether further operations are required. The vibrating beam tamper can be single or double (a twin beam compactor). The beam is lifted and moved forward in increments that should not exceed the width of the beam, which is usually 75–100 mm. Poker vibrators should be used around the perimeter to help ensure adequate compaction.

The concrete between the side forms should be finished with a surcharge of about 50 mm to allow for reduction due to compaction. The actual surcharge will depend on the thickness of the slab and the compacting factor of the concrete. If the surface is left as it comes from the tamper, it will be fairly rough, which may be needed in multi-storey car parks but may be rather too rough for other purposes. To achieve a smoother finish, the tamper should be taken back every 1.5 to 2.0 m and then moved forward slowly over the compacted surface to smooth out the ridges and furrows left by the first pass of the beam. The floor surface finish can be further improved by the use of a scraping straight-edge.

Another method is to use a power float, followed by a power trowel as a follow-up to the tamping (Fig. 3.2). To achieve satisfactory results, the concrete must have reached the right degree of stiffness before the power floating is started. This requires considerable experience on the part of the operators (Fig. 3.3).

An alternative to power floating and trowelling is to wet grind the surface after the use of the scraping straight-edge. The wet grinding can also be used after completion of the power floating. Wet grinding

Fig. 3.2 View of an industrial floor receiving power trowelling. Courtesy: Monk Construction Ltd.

Fig. 3.3 View of the 'Malconcrete Process' floor for large warehouses. Courtesy: Hughes Group Holdings PLC.

should not take place for at least 14 days after completion of the floor. Wet grinding will provide an excellent surface for the application of surface sealants, and is likely to be needed to achieve the highest standard of 'flatness', i.e. a 'super-flat floor.

'Dusting' of concrete floor surfaces – sealers and hardeners

It is claimed, quite correctly, that good-quality concrete, properly laid, compacted, finished and cured, does not normally require any further surface treatment. However, the author does recommend the application of a good-quality surface sealant such as Lithurin (magnesium-silico-fluoride); this is not expensive and is very effective in controlling 'dusting' which is bound to occur to some extent to any concrete floor under the action of traffic (foot and machine). It combines with the free lime in the concrete and forms harder more durable compounds.

Sealants based on single pack, moisture cured, polyurethane systems also give very good results.

With all types of surface sealants it is essential that the compound used should penetrate the surfaces layers of the concrete; a penetration of 10 mm is often claimed by manufacturers, but with high-quality concrete this is rather optimistic.

There are, of course, a large number of proprietary compounds on the market sold under the general name of 'concrete sealers' and 'concrete hardeners'. The former are applied to the surface of the concrete after completion of all finishing operations; the latter are generally, but not always, added to the concrete mix in the mixer.

Many surface sealers are claimed by the producers also to harden the surface of the concrete and thus increase its abrasion resistance. The author's experience is that provided the completed floor slab is good quality and the sealer/hardener is correctly applied in accordance with the manufacturers' instructions, abrasion resistance and resistance to the penetration of oils and non-aggressive liquids (such as water), are increased.

Sprinkled finishes

All sprinkled finishes are proprietary materials; some are non-oxidizing metallic, chemically-inert fine aggregate, and others are chemically-inert non-metallic fine aggregate.

The objective in using them is to provide a harder and more abrasion-resistant surface than would be obtained by the use of the ordinary high-strength concrete base slab. Suppliers claim, typically, that the rate of wear is only about one-eighth of that on a high-strength power-floated concrete.

These materials are supplied in various grades designed to meet the anticipated use to which the floor will be put. Generally, this type of finish is suitable for medium, heavy and very heavy duty floors, particularly to resist impact. All these sprinkle type finishes are laid monolithically with the base concrete, so that the correct time for the application must be carefully gauged by the applicators. It is usual for the base concrete to be allowed to stiffen until it is suitable for power trowelling.

Sprinkled finishes can of course be applied to toppings and over-slabs. Some suppliers provide the aggregate prepacked with cement so that it is ready to use; an example is 'Supertop' by Armorex Ltd, Bury St Edmunds, which provides a trowelled-in thickness of about 3 mm (Fig. 3.4). On the other hand, the proprietary 'Trowelling-in-Betonac' aggregate, supplied by Don Construction Chemicals Ltd, Doncaster, is recommended to have a trowelled-in thickness of about 4 mm.

Fig. 3.4 View of turbine hall floors at Sizewell B Nuclear Power Station finished with Armorex Supertop dry shake, and Proseal dustproofer. Courtesy: Nuclear Electric and Armorex Ltd.

Screeds

Introduction

Screeds have been defined earlier in this chapter, but the author wishes to emphasize that in this book screeds are not toppings, and are intended to form an intermediate layer between the structural slab and the flooring, and therefore are not suitable to form the final wearing surface of the floor. This 'intermediate layer' may be needed to make up levels, and/or to provide a smoother level surface (than that provided by the concrete deck) on which to lay the flooring.

Cement/sand screeds

Screeds are generally made with ordinary Portland cement, sand and water. Some mixes are 'modified' by the addition of an acrylic resin or a styrene butadiene rubber (SBR) emulsion, or other suitable polymer. A later section in this chapter deals with these polymer-modified screeds.

There are misconceptions about the need for cement-rich mixes for screeds. While the amount of cement in the mix is important, the water/cement ratio, method of mixing, workability, degree of compaction and standard of curing are also important, and have a profound effect on the durability of the screed and its 'fitness for purpose'.

The main Code of Practice that makes recommendations for floor screeds is BS 8204: In-situ floorings: Part 1: Code of Practice for concrete bases and screeds to receive in-situ floorings.

Useful and practical recommendations are also given in BS 8000: Workmanship on Building Sites: Part 9: Code of Practice for cement-sand floor screeds and concrete floor toppings.

Materials, batching, mixing, compacting and curing

The author's experience indicates the following:

- Mix proportions: 1:4.0 to 1:4.5 by mass (weight) assuming a dry sand.
- Volume batching can be accepted for small floor areas based on one bag of cement (assumed to be 35 litres) to about 120 litres of sand. Batching by volume results in greater variation in the actual mix proportions and hence wider variations in the quality of the screed.
- Water/cement ratio: 0.4 (20 litres of water to 50 kg cement, assuming a dry sand); the moisture content of the sand must be taken into account. See the comment below regarding the importance of the grading of the sand.
- Mixing should *not* be in a free-fall concrete mixer, but in a pan paddle mixer or forced-action mixer. Careful hand mixing using either gauge boxes or buckets (but not shovels) for obtaining the prescribed mix proportions, can be used for small floor areas.

Sand

The sand should be clean, concreting sand to BS 882 (latest edition), grading limits C or M. The grading of the sand is of great importance in obtaining adequate workability for proper compaction and this is a key factor in the batching of screed mortars by firms supplying ready-mixed mortars. Such a degree of precision is unlikely to be obtained with on-site mixing and in such cases, the use of an SBR emulsion as a workability aid should be considered.

Placing
The mortar can be placed by manual methods or by pumping. The latter is generally favoured for large floor areas, but special attention is needed to ensure that the specified mix proportions are maintained, or only modified with approval of the specifier's representative. Special pumps are used and this enables a fairly stiff and cohesive, but workable, mix to be used.

A water/cement ratio of 0.4 should generally not be exceeded.

Compaction
The mortar should be thoroughly compacted and the oven-dry density of the hardened screed should not be less than about 1900 kg/m^3.

Curing
Curing should last for 3 days and consist of covering the freshly-laid screed with polythene sheets well lapped and held down around the perimeter by blocks or scaffold boards. A curing membrane should only be used if it is not intended to bond the floor covering to the screed.

Laying criteria

General considerations
Screeds can be fully bonded, partially bonded or unbonded, to the base concrete. To achieve a 'fully bonded' 'condition, the screed should be laid monolithically with the base concrete – a method very seldom used.

Partial bonding is satisfactory for the majority of floor screeds and can be achieved by thoroughly cleaning the surface of the base slab and damping down prior to laying the screed. The use of a SBR/cement slurry (50 kg cement to about 30 litres of SBR), will help to ensure good bonding, but it is essential that the mortar be laid on the bonding coat while the latter is still tacky. While an 'average' of 30 litres per 50 kg (35 litres) of cement is reasonable, the slurry should be of 'brushable' consistency and should be applied while the concrete surface is damp, but not wet; any pools of water must be removed.

Bush hammering or grit blasting can be used when a particularly good bond is required, but all dust and grit must be removed before the bonding layer or screed mortar is laid on the concrete. Equipment that combines grit blasting with a vacuum and dust and grit collection device is readily available.

If the bonding layer has dried out it will form an excellent *debonding* layer. The bonding aid must be used in accordance with the manufacturer's instructions.

The author does not consider that a bonding aid can be accepted as an effective damp-proof membrane (dpm) in ground floor slabs, and recommends that the dpm be laid as a separate operation. Unless special precautions are taken, the dpm is likely to debond the screed from the base concrete. The 'special precautions' will depend on the type of dpm used. With epoxies, a special primer can be used on the cured epoxy coating; with rubber-bitumens, coarse sand can be sprinkled on the dpm while it is still tacky, but some reduction in bond (compared with laying the screed directly on the concrete) must be accepted.

Bay sizes and layout

The author favours the laying of floor screeds in bays between screeding boards, the width of the bays should not exceed about 3.0 m. This should enable the screed material to be properly compacted.

The length of the bays is less important than the width as the fine transverse cracks that will form are unlikely to affect adversely the durability of the screed.

The trouble with joints is that they tend to show through thin flexible flooring materials such as vinyl sheet and tiles, whereas fine cracks should not do so. For this reason proprietary screeding materials have been developed and are on the market. These are self-levelling mortars with small drying shrinkage and rapid hardening characteristics. Some information on these special materials is given later in this chapter.

Compaction

An adequate surcharge should be provided above the side screeding boards (about 10 mm is usually sufficient). Compaction must be thorough in order to achieve adequate impact resistance and good bonding with the base concrete. The thickness of the screed will be determined largely by the bonding conditions to the base concrete.

With 'fully bonded' (monolithic) construction, the screed need not be thicker than 15 mm; with partial bonding, a minimum thickness of about 30 mm is recommended; while for unbonded conditions the minimum thickness should be 50 mm.

Unbonded conditions would apply when the screed is laid on a damp-proof sheet-type membrane, or an in-situ membrane, unless the precautions mentioned above have been taken.

In refurbishment contracts, levels sometimes vary widely and then consideration should be given to the desirable maximum thickness for mortar screeds. The author recommends a maximum thickness of 75 mm. If a greater thickness is required then it is generally better to use a fine concrete (10 mm aggregate). The thicker the screed, the greater is the compactive effort needed to ensure proper compaction. The increase in depth increases the drying shrinkage 'gradient' through the screed, and this increases the risk of cracking, and curling at bay joints.

Screeds laid on thermal insulation

Due to the upgrading of thermal insulation requirements for most types of buildings, it is becoming increasingly common for floors to incorporate some form of thermal insulation. Reference should be made to the Building Regulations 1985 and Approved Document L, Conservation of Fuel and Power (1990). This document introduced for the first time requirements for U-values for ground floors of 'dwellings and all other buildings'. This is given in Table 1 of the document as a maximum of 0.45 W/m^2K. Reference should also be made to Building Research Information Paper No.IP.3/90 of April 1990 – The U-value of ground floors: Application to the building regulations.

The above documents can be summarized by saying that if the perimeter-to-area ratio does not exceed about 0.25, then special thermal insulation will not be required, and this applies to all types of ground floors, i.e. slab on the ground, suspended concrete, and timber floors.

However, the above are the minimum legal requirements, and to conserve fuel, lower U-values are often adopted; this requires the use of thermal insulation in the floor system. The usual location for the insulation is between the floor slab and the screed.

The author strongly recommends the use of a relatively incompressible insulation such as extruded polyurethane, rather than a compressible quilt. The screed is rigid and comparatively brittle when carrying floor loading and is liable to crack and may break up if the insulation on which it is laid is not reasonably rigid. When serious failure does occur the only effective remedy is to remove both screed and quilt, and re-lay the screed on a rigid type of insulation.

Even when extruded polyurethane is used, the inclusion of a light mesh reinforcement (not chicken wire) is recommended, as this will help to hold the screed together after it has cracked, as some degree of cracking is inevitable. In ground floor slabs it is important that the thermal insulation is of a type that does not absorb water as this reduces its insulating properties; in any case it is prudent to place the thermal insulation above the damp-proof membrane, which in floor slabs in contact with the ground is required by the Building Regulations 1991 and Approved Document C: Site preparation and resistance to moisture: 1992.

Polymer-modified cement/sand screeds

In the UK acrylics and SBRs are the most commonly used polymers in cement/sand mortars for screeds. When properly used they can impart the following properties to the screed mortar:

– improved workability with a constant water/cement ratio;
– lower water/cement ratio with constant workability;

– reduced water absorption;
– improved chemical resistance;
– improved bond with the substrate.

The author's experience mainly relates to the SBRs and he can confirm that when correctly used the improvements listed above are in fact achieved.

The proportions of cement to sand should be 1:4.0 to 1:4.5 by mass. The sand should be clean and comply with the grading limits C or M in Table 5 of BS 882: 1983 (or a subsequent revision).

An excellent paper on the properties of latex-modified Portland cement mortars by D G Walters was published in the *Materials Journal of the American Concrete Institute* (July–August 1990), which describes tests on five types of latex used as admixtures for cement/sand mortar mixes:

(a) plasticized polyvinyl homopolymer (PVA);
(b) a copolymer of vinyl acetate and ethylene (VAE);
(c) a carboxylated styrene butyl acrylate copolymer (S-A);
(d) a carboxylated butyl acrylate – methyl methacrylate copolymer (PAE);
(e) a carboxylated styrene butadiene copolymer (S-B).

The paper compares the five latex mortars for water/cement ratio, permeability, adhesion, compressive strength, flexural strength, weathering resistance, freeze-thaw resistance, acid resistance and carbonation resistance. While some of the properties investigated are not really applicable to floor screeds, the overall results are interesting and significant. The author concludes that the polyvinyl acetate latex should not be used where there is any chance that the mortar will be exposed to moisture, and that the properties of the PVA modified mortar are generally significantly poorer than those of the other latexes. The styrene butadiene copolymer modified mortar gave the best results in almost all of the tests.

Proprietary early-drying screeds

There are early-drying screeds now on the market and the principal claims made for them are that they are:

(a) quick setting and rapid hardening,
(b) early drying and with low shrinkage characteristics,
(c) of high strength and resistance to impact,
(d) easy to lay; some are virtually self-levelling.

These special properties appear to be due to the 'unique' characteristics of the cementing material used in the mix which locks in a very high percentage of the gauging water, thus reducing moisture loss and the resulting drying shrinkage. These proprietary materials cost considerably more per square metre than ordinary cement/sand screeds including polymer-modified mixes, but the advantages outlined above will often more than compensate for the extra cost.

Time is always an important factor in construction and floor finishes are always applied towards the end of the contract when there is great pressure to reach 'practical completion'. Ardex UK Limited, Haverhill claim that screeds made with their 'Ardurapid 35 Cement' will pass the BRE screed impact test in about 5–6 h after laying. For normal cement/sand mixes, a minimum of 14 days is required. Similar claims are made for the Flowcrete ED Screed.

Suppliers of flooring (floor coverings) have strict requirements for the maximum moisture content of the substrate on which the flooring will be laid. These requirements give rise to many disputes on site, and so it is important to state clearly in the contract documents what the requirement is and how it is to be measured. This is discussed in some detail under the section dealing with sheet and tile flooring, where reference is made to BS 8203: Code of Practice for the Installation of sheet and tile flooring, in Chapter 4.

As these new materials are all proprietary it is neither practical nor desirable to make detailed recommendations for laying. All that can be said when including such materials in a contract specification or considering a request by a contractor to use one of these materials is that:

(i) the standing of the supplier should be checked, and relevant claims made for the material's characteristics should be checked in some detail;
(ii) jobs where the material has already been used should be investigated;
(iii) approval should be subject to strict compliance with the supplier's instructions and to some degree of site control by the supplier's technical representative.

Testing floor screeds

It is unfortunate that floors give rise to more complaints and disputes than almost any other part of a building, with the possible exception of flat concrete roofs. The basis of such complaints is that the completed screed is not 'fit for the purpose'. This is linked with the allegation that the screed does not comply with the contract and/or that it does not comply with the recommendations in the relevant British Standard Code of Practice.

Until the early 1980s testing consisted mainly on determining the cement content, which was often found to be below that required by the mix proportions in the contract (usually a cement: sand ratio of 1:3 without reference to volume or weight). The real faults were usually inadequate preparation of the base concrete, totally inadequate compaction (due to the use of too dry a mix), the use of sand with too fine a grading, and inadequate or complete absence of curing.

At the end of the 1970s the Building Research Establishment developed an impact screed tester which has revolutionized the site testing on floor screeds. This equipment is not entirely suitable for testing screeds laid on thermal insulation. Reference can be made to BRE Information Paper IP11/84. More detailed comments and brief recommendations on investigations and testing are given in Chapter 6.

Toppings

Introduction

A definition of topping is given at the beginning of this chapter. Toppings are usually based on Portland cement (high alumina cement for emergency work), polymer resins, bituminous emulsions, or magnesite. The intention is to form a hard-wearing surface which may be required to possess special characteristics to suit the particular use to which the floor will be put.

In-situ concrete toppings

General considerations

In-situ concrete toppings are bonded to the base concrete and may be laid *monolithically* with the base slab, or as *separate construction*, i.e. laid some time after the slab has hardened. The previous Code of Practice, CP 204, contained recommendations for thicknesses of both monolithic and separately-constructed toppings, but these recommendations have been omitted from the new Standard, BS 8204: Part 2. The author feels that some guidance is needed on minimum thicknesses and on acceptable tolerances for the thickness, and these are given below:

- Monolithic construction: Nominal thickness, 20 mm
                             Tolerance, –10 mm
- Separate construction: Nominal thickness, 40 mm
                             Tolerance, –20 mm

It will be noted that in relation to the minimum thickness, the tolerance for monolithic toppings is 50%, and for separate construction it is also 50%. The reason for this is that the monolithic topping is laid on semi-plastic concrete and therefore it is not practical to achieve a tight tolerance on thickness. It is not intended that the tolerances given above should be interpreted to indicate that over a high percentage of the area the topping need only be 10 mm or 20 mm thick. The author considers that the relevant minimum should be achieved over 75% of the area. In other words not more than 25% of cores taken, or measurements made before the topping has hardened, should show that the topping is 10 mm or 20 mm thick, provided the measurements taken are really representative of the floor as a whole.

The topping can be either high-quality concrete on its own or the aggregate can be specially selected, e.g. granite, or the surface of the topping can receive a sprinkle finish of metallic/synthetic aggregate, applied and trowelled in while the concrete is still plastic.

Monolithic construction

From the point of view of obtaining maximum bond with the base slab, monolithic construction is undoubtedly the best construction, but it does introduce considerable practical difficulties on site. The topping mix has to be laid while the base concrete is still 'green', i.e. it still retains some plasticity. The time between completion of placing and compacting the base concrete and when the concrete has stiffened sufficiently to allow the topping to be laid will depend mainly on the mix proportions and the ambient temperature, and this can be anything from 1 to 4 h. It will be seen that night work is inevitable with resulting problems of site control and supervision. For this reason, monolithic construction is seldom used. The tolerances on surface finish and the wearing quality of the topping are sometimes disappointing, although the bond with base slab is usually excellent.

If the topping is judged unacceptable, e.g. well outside the specified tolerances and/or has poor abrasion resistance, then great problems can arise from attempts to remove the topping as it is monolithic with the base concrete. The author knows of cases where attempts to remove a very unsatisfactory monolithic topping have had to be abandoned and a different solution found for the remedial work. Problems of remedial work are discussed in Chapter 7.

Separate construction

When a topping is laid on a hardened concrete base slab, the process is known as 'separate construction' even though all reasonable steps are taken to ensure good bond at the interface.

There are differences of opinion among professional floor layers over what should be done to ensure 'adequate' bond. This raises the question of what is 'adequate' bond, and this is not an easy question to answer.

Shortcomings in bond at the interface show as curling at the perimeter of bays and as hollow-sounding areas within the bays, known collectively as 'lack of adhesion to the base'. Clause 121 in CP 204 stated: 'Loss of adhesion does not necessarily mean that the screed or concrete topping is unsatisfactory. However when it is accompanied by visible or measurable lifting of the edges of bays or at cracks, the screed or concrete topping may defect or break under the loads imposed in use. . . .'

This part of CP 204 was withdrawn in 1987 and replaced by BS 8204: Part 1.

The author feels that the standard of bond required should be related to the use to which the floor will be put, and to the location of the defective areas. For example, lack of adhesion (debonding) of a screed or topping immediately adjacent to a wall is much less likely to cause trouble than in a main traffic route. The standard needed for domestic and lightly-loaded commercial use can be lower than that required for heavy commercial and industrial use, where impact will be severe.

In all cases the surface of the base slab should be carefully examined for signs of contamination by oil, grease, etc., as the presence of such contaminants will have an adverse effect on bond.

To secure the best bond, the coarse aggregate in the base slab should be exposed, all grit and dust removed and the surface well damped down the day before the topping is laid. A suitable bond coat can then be applied immediately before the topping is laid, but care must be taken to ensure that there is no standing water on the surface of the concrete slab (Fig. 3.5).

The strength/quality of the base slab must be compatible with that of the topping; a cement-rich high-strength topping may debond itself from a weak base slab.

In cases where maximum bond is not considered necessary and the base slab is good quality, clean and uncontaminated, the exposure of the coarse aggregate can be omitted and reliance placed on the use of a suitable bond coat, provided the bond coat is applied strictly in accordance with the directions of the suppliers.

Materials and construction

In the past, the standard high-strength in-situ concrete topping was granolithic. With the revision of Code of Practice for in-situ floor finishes (CP 204, Section 2) in 1987 and the publication of BS 8204: Part 2, the term 'granolithic topping' was replaced by 'High-strength concrete toppings' (including granolithic).

Table 2 in BS 8204: Part 2: 1987 gives recommendations for the grade of concrete, minimum cement content, type of fine and coarse aggregate, which are likely to be satisfactory for four standards of use,

Fig. 3.5 A well-scabbled concrete floor surface prepared to receive a high-strength concrete topping.

from severe abrasion and impact to moderate abrasion and light industrial and commercial traffic.

While the author is a firm believer in adequate cement contents, he has reservations about the need/desirability of using 475 kg of cement per cubic metre for any in-situ concrete floor slab for industrial use.

Proprietary cement-based toppings and finishes

There are a number of proprietary pre-packed materials on the market that can be used instead of granite aggregate concrete topping to provide an exceptionally hard-wearing floor. They should all be used exactly in accordance with the directions of the suppliers. When specifying such material, it is most desirable that responsibility should be clearly defined and the suppliers of the material should be required to undertake site visits at predetermined intervals, and to provide written reports on each visit with copies to the supervising officer/resident engineer, main contractor and sub-contractor.

A typical example of this type of topping is 'Diamond Betonac' which can be laid as a monolithic topping with a 12 mm minimum thickness or as 'separate construction' with a total minimum thickness of 37 mm.

The 37 mm is made up of what the suppliers call a bonding layer composed of 1:1½:2½ concrete using 10 mm nominal size aggregate, with a minimum thickness of 25 mm, followed by the 12 mm Betonac topping laid monolithically with the bonding layer. The base concrete has to be thoroughly hacked to expose the coarse aggregate and all dust and loose particles removed. The author recommends that for this type of topping the existing concrete base slab should have a minimum compressive strength of 25 N/mm^2.

The bonding layer is 'cement rich' as it contains about 430 kg of cement per cubic metre, which in the opinion of the author is rather high. A compressive strength of 30 N/mm^2 would appear to be adequate, taking into account the 12 mm Betonac topping.

An example of a polymer-modified cement-based topping is 'Nitoflor Cemtop'. It can be used to provide a wearing surface on new concrete floors and for providing a new wearing surface on an old deteriorated concrete floor. The Nitoflor System consists of four items: a primer (Nitoprime 33), a base layer having a minimum thickness of 8 mm (Cemtop Base), a second priming coat, and final wearing surface of Cemtop HD (a minimum of 7 mm thick). Very high abrasion and impact resistance and the ability to be laid in large areas without shrinkage cracking and/or curling (debonding) are claimed for this topping. The base concrete must have a minimum compressive strength of 20 N/mm^2 and the surface must be prepared by lightly exposing the coarse aggregate and removing all loose grit and dust.

Proprietary resin-based toppings and finishes

When using resin-based types of materials, especially epoxies, the Health and Safety at Work Regulations may require special precautions to be taken. The principal resins in use in the UK are epoxies, polyurethanes and polyester; the last of these being used appreciably less than the first two.

The thickness varies between 2 and 12 mm, but is mainly in the range 4–6 mm. The systems include the use of selected aggregates and fillers.

The 2 mm thick toppings are often self-levelling. An example is the NITOFLOR SL2000 and SL4000. A variation of this type of topping is to provide anti-static characteristics; the whole of the topping exhibits high electrical conductivity and so when correctly earthed will be electrically conductive and prevent the build-up of undesirable static electricity. The thicker toppings are applied by trowel to the required thickness. The basic resin is epoxide and the toppings are abrasion and impact resistant and also resistant to a wide range of chemicals. When specific chemical resistance is required, reference must be made to the suppliers. In general, resins are vulnerable to organic solvents.

A polyurethane-based industrial floor topping is 'Ucrete' which was developed and patented by ICI in the early 1970s and is now made and marketed by Thoro System Products Ltd, Redditch, which is part of ICI Specialities. It is marketed for industrial use in seven grades, including slip resistant and anti-static grades (see Fig. 3.6).

Fig. 3.6 View a floor in a food factory surfaced with 'Ucrete'. Courtesy: Thoro System Products Ltd.

The recommended minimum thickness for heavy industrial use is 9 mm and for lighter use 4–6 mm. Ucrete is a bonded topping and the detailed directions of the manufacturers should be carefully followed.

Ucrete can withstand temperatures up to 100°C and so very hot water and steam can be used for cleaning. The coefficient of thermal expansion is claimed to be similar to that of Portland cement concrete. The topping has good impact and chemical resistance, but its resistance to specific chemicals should be checked with the manufacturers. It has low water permeability and absorption.

Bituminous-based toppings

Bituminous-based toppings are composed essentially of bituminous emulsions, Portland cement, sand and chippings. These toppings are all proprietary and can be considered as semi-rigid. This type of material forms a virtually jointless, dustless, resilient and quiet surface with a probable 'life' of about 20 years. Fine cracks are normally self-sealing under traffic; wider cracks and damaged areas can be readily repaired.

The basic types are all black but if this is considered unacceptable, a more decorative colour finish can be supplied which is generally a resin-based material containing silica sand and fillers. A material of this type which has been on the market for many years is 'Latexfalt' by Amey Flooring Ltd, Abingdon, and another is Mastic Flooring by Colas Building Products Ltd, Chester.

The Colas product is normally laid on a prepared concrete base slab to a compacted thickness of 12 mm. The concrete is not scabbled, but the surface must be free of grease, oil and other contaminants, and be well washed with water and allowed to dry. A priming coat is applied consisting of bitumen emulsion diluted with an equal quantity of water, and allowed to dry. On this, a bond coat of undiluted bitumen emulsion is applied and must be still tacky (not dried out) when the mastic is laid on it. The mastic is spread, levelled and compacted either by manual methods or by mechanical equipment. Final compaction of the Mastic Flooring is achieved by rolling both before it has cured (24–36 h) with a 136–226 kg roller, and after curing (about 48 h) by means of a hand-controlled vibrating roller. The floor can be used by foot and light vehicular traffic some 36 h after laying, and heavy traffic some 72 h after laying. The suppliers provide detailed instructions for laying their Mastic Flooring and all necessary technical information.

In-situ terrazzo toppings

At the time of writing this book, formal recommendations for in-situ terrazzo toppings are given in Section 3 of Code of Practice, CP 204. The Code is supplemented by a Specification of the National Federation of Terrazzo-Marble and Mosaic Specialists.

The aggregate is selected marble chippings, usually with a maximum size of 10 mm free from dust and elongated/flaky particles. If large aggregate is used, the thickness of the topping should be correspondingly increased over the normal thickness of 15 mm. Usually white Portland cement is used in preference to OPC which can vary in colour from light to dark grey. The topping can be laid on a cement/sand screed or directly onto a prepared concrete base slab. The wording of the Code suggests that it is better to lay the topping on a separating layer in the case of suspended floors. The author's experience is that separating layers are best avoided unless there is a good reason to use them, e.g. when there will be an appreciable thermal gradient through the floor. If a separating layer is used, the concrete slab on which it is laid should be provided with a power-floated or hand-trowelled finish. A tamped finish should not be accepted.

It is 'trade' practice to use a very rich mix, with a cement to sand ratio of 1:2 by volume. This is about 1:2.3 by mass which, with a water/cement ratio of 0.4, contains about 560 kg cement/m^3. High drying-shrinkage stresses are generated and this will inevitably result in cracking unless the following precautions are taken, and even then complete success cannot be guaranteed:

(a) A maximum w/c ratio of 0.4 (preferably lower) which may require the use of a plasticizer in the mix.
(b) The marble aggregate should be virtually dust free; the aggregate particles should not be less than 3 mm in diameter, and should not be flaky.
(c) The terrazzo should be laid in bays of 1 m × 1 m with a dividing strip between. If specially shaped bays are required for visual effect, each bay should not exceed 1 m^2 in area, and the length should, theoretically, not exceed 1.5 times the width, although it is normal practice to accept a factor of 2.0; in the opinion of the author, this increases the risk of formation of transverse cracks.
(d) Curing should be for at least 4 days by covering with polythene sheeting, placed in position as soon as it will not mark the surface. Screens should be provided if there is danger of draughts impinging on the newly-laid terrazzo. To reduce the risk of crazing, the author recommends that the greatest care be taken to prevent the rapid drying out of the surface of the newly laid terrazzo for the first 48 h. Protection of the surface should be provided within minutes of completion of the final trowelling.

The substrate on which the terrazzo is laid must be clean, and well damped down prior to commencement of laying. Thorough compaction by tamping, trowelling and rolling is essential for strength and durability.

The surface is ground and polished, but this should not be commenced until at least 6 days after completion of laying. In-situ terrazzo, properly laid, provides a most attractive floor finish, but as mentioned above, the great danger is drying-shrinkage cracking. This can take the form of 'normal' shrinkage cracks which may extend down through the full depth of the terrazzo, and/or crazing which is a fine network of cracks which seldom exceed 3 mm in depth; they form a pattern rather like chicken wire, enclosing contiguous areas having dimensions in the range of 15–50 mm across. Crazing occurs in the very early life of the terrazzo and may not be visible for some time, i.e. until the surface is wetted or the fine cracks become filled with dirt. The presence of crazing can detract significantly from the appearance of the terrazzo, but is not structurally serious and ordinarily does not indicate future deterioration. Some additional information on crazing is given below.

Cleaning can best be carried out by using warm water with a neutral detergent. Stains and resistant dirt can usually be removed by the careful use of a fine abrasive powder. Soap should not be used as this will make the floor slippery.

Crazing of concrete surfaces

Crazing of the surface of an otherwise well laid in-situ terrazzo floor topping can turn into a disaster for the contractor concerned. It is most unlikely to have any long-term effect on the durability of the floor, but because dirt gets into the fine cracks, and as terrazzo is a highly decorative material, it can spoil the appearance of the floor.

The phenomenon of surface crazing is not confined to in-situ terrazzo, but can occur on any concrete surface, and this is emphasized in reports by the Cement and Concrete Association (see the Bibliography at the end of this chapter). It is often claimed that concrete made with white cement is more prone to crazing than concrete made with grey cement. There are differences of opinion on this, as the dirt that gets into the fine cracks is more noticeable on a white surface than on a grey one. However, work by Levitt for the Cast Concrete Products Industry, does indicate that grey cement has a lower drying shrinkage and wetting expansion than white cement. The Cement and Concrete Association carried out research on 94 concrete columns cast in formwork, over a period of some 18 months. This and other work by the Cement and Concrete Association and others indicated that crazing was essentially the result of surface tensile stresses caused by the shrinkage of the surface relative to the mass of concrete below the surface; also, that this shrinkage could be attributed to the following:

(a) differential thermal movements;
(b) differential moisture movements;
(c) carbonation of the surface.

F.M. Lea in his internationally known book, *The Chemistry of Cement and Concrete*, (1970) third edition, Edward Arnold, clearly favours carbonation of the surface as a major factor. The Cement and Concrete Association conclude that '. . . it seems unlikely that carbonation shrinkage on its own leads to crazing, but rather that the brittle layer formed by carbonation of the surface crazes as a result of both drying and carbonation shrinkage'.

(d) Where concrete is cast within formwork, the type of formwork surface in contact with the concrete is also important.

A perusal of the extensive literature on this subject indicates that the four factors mentioned above all play their part in the formation of crazing, but their relative importance depends on whether the concrete is cast within formwork (as was the case with the Cement and Concrete Association columns) or whether the concrete is cast as a slab (e.g. in-situ terrazzo).

In the case of slabs, the two factors (a) and (b) above, are greatly influenced by the details of the mix and standard of curing, including the protection of the surface immediately following final trowelling. Recommendation for reducing crazing include the following:

(1) Mixes rich in cement should be avoided.
(2) The lowest practical water/cement ratio should be used.
(3) The practice of trowelling in dry cement or a mixture of cement and crushed rock fines to absorb bleed water should not be followed.
(4) Proper curing and protection of the surface against sun and wind should start immediately after the completion of the final trowelling.

It is obvious that the above will not prevent the surface carbonating, but as mentioned above, authoritative opinion is that carbonation shrinkage alone is unlikely to result in *visible* crazing.

Magnesite (magnesium oxychloride) toppings

At the present time there is no Code of Practice for laying magnesite floor toppings, as Section 7 of CP 204 was withdrawn in 1991. The reason for withdrawal was that the Code recommendations were out of date and not in accordance with current floor-laying practice. Also, over 90% of magnesite flooring in the UK was laid by only two contractors and so it was in reality a proprietary material and not suitable for inclusion in a Code of Practice.

There is, however, a British Standard, BS 776, for materials for magnesium oxychloride (magnesite) flooring. This includes requirements for quality, colour, composition, fineness, setting time, strength and linear change, of the calcined magnesite; also requirements for the magnesium chloride, wood flour, sawdust, aggregates, fillers and pigments.

Thus as long as BS 776 remains valid, there are clear quality requirements for the materials used in the laying of magnesite flooring. For these requirements to be effective, they must be incorporated in any contract for the laying of this type of floor topping. The specialist contractors are free to select the materials listed in BS 776 and proportion and mix them as they think fit.

The author's experience indicates that the following general comments are applicable:

(a) The topping constitutes a 'breathing' floor which means that water vapour can pass through it (the same applies to concrete toppings and cement-based screeds).

(b) The material is vulnerable to trapped moisture and wet conditions. The author would not recommend the use of a magnesite topping for a 'wet' trade, nor for a ground floor slab where there was a high water table, unless an effective damp-proof membrane was included in the floor construction.

(c) It is not advisable for a magnesite topping to be sealed with a vapour-resistant membrane.

A properly laid magnesite topping is hard wearing and 'warm' to the feet. It is laid in large areas with the minimum of joints; any cracks that may develop due to stresses in the base slab can be readily repaired. The topping is not subject to drying shrinkage (Fig. 3.7).

Fig. 3.7 View of an industrial floor finished with 'Quil-Magna' magnesite topping. Courtesy: A. Quiligotti & Co. PLC.

A 'typical' mix (by mass) for a magnesite topping would be:

- 1 part magnesite (magnesium oxychloride),
- 1 to 3 parts of filler (which includes fine aggregate),
- a small quantity of wood flour and pigment, and
- a solution of magnesium chloride as the gauging liquid.

The actual mix proportions depend on the type of use (light or heavy) for which the floor is required.

Overslabbing with mastic asphalt

The relevant British Standard is BS 6925: Mastic asphalt for building and civil engineering (limestone aggregate), which replaces BS 988, BS 1076, BS 1097 and BS 1451. The Standard includes requirements for flooring grades of mastic asphalt composed of ground limestone, coarse aggregate, asphaltic cements, and pigment if required; appendices give recommendations for grades and thicknesses for flooring.

Mastic asphalt as an overlay would be used mainly for industrial floors, but is also used on external balconies, access decks and podia. When laid as an overlay on a concrete floor slab, the normal thickness would be about 30 mm, and it would be laid on a glass-fibre tissue separating membrane.

The asphalt is usually given a sand-rub finish. If special skid resistance is required, a polymer resin-bonded grit can be applied to the surface. Calcined bauxite is often used, as this has the great advantage that it does not 'polish' under traffic.

Mastic asphalt is not suitable where large-scale spillage of petroleum products is anticipated; protective coatings of epoxy or polyurethane resins can be applied in the spillage areas. If the overslab is required for a floor that has to be watertight, then the paving grade asphalt (BS 1447) should be laid on 20 mm of roofing grade asphalt (BS 6925) in two layers of approximately equal thickness. The roofing grade asphalt should be laid on a glass-fibre isolating membrane. For watertight work, a skirting of two-coat roofing grade asphalt to a total thickness of about 13 mm, and 150 mm high should be provided, but the actual height of the skirting will depend the circumstances of each case.

It is recommended that the paving-grade mastic asphalt should be laid on the roofing grade as soon as possible to help ensure a reasonable bond between the two types of asphalt.

Overslabbing with in-situ concrete

Overslabbing is defined at the beginning of this chapter. For the purpose of this section it is assumed that the in-situ concrete overslab is structurally separated from the existing floor slab by a satisfactory separating membrane, such as 1000 gauge polythene sheeting or two layers of say 500 gauge sheeting. In other words, it is a new floor slab. This will of course increase significantly the dead load of the floor which can have serious consequences, particularly for suspended slabs. If it is desired to limit the increase in dead load, consideration can be given to the use of structural lightweight aggregate concrete.

The usually-accepted minimum thickness of in-situ concrete over-slabs is 100 mm and frequently it is appreciably more, depending on such factors as:

(a) the condition, both physical and structural, of the existing floor;
(b) the use to which the floor will be put;
(c) the floor levels of adjoining parts of the building; and
(d) whether the existing floor is 'uniformly supported' on the ground or whether it is suspended.

The specification/design of the overslab will have to take into account the above factors, with special reference to (b) and (d).

Floors and roofs for car parking

Floors and roofs used for car parking present a number of special problems which, excluding structural design, are summarized below:

(a) Both floors and roof must be watertight.
(b) The running surfaces must be reasonably even but possess non-skid/good grip characteristics, which should be durable.
(c) Detailing and construction of joints, particularly movement joints, require special care.
(d) If ceiling finishes to the soffits of roof and floor slabs are not provided (which is usually the case), care should be taken in the specification and execution of 'off-the-form' concrete finishes.

Watertightness

Water passing through concrete becomes highly alkaline and if this comes into contact with cars parked below, damage is caused to the paint.

The use of post-tensioned slabs can provide a solution to the problem of thermal and shrinkage cracking; also the use of high-quality, dense, low-permeability concrete, and if considered desirable, the application of a low-viscosity polyurethane sealant on the top (running) surface of intermediate decks, or some other relatively-thin, abrasion-resistant, crack-free, self-levelling, polymer-based topping.

Roofs must either be finished with mastic asphalt or with a proprietary polymeric coating that is durable, resistant to traffic, can accommodate movement, and bonds well to the concrete base slab. Proprietary materials must be applied in accordance with the recommendations of the suppliers, and carried out by specialist contractors.

Figure 3.8 shows a car-park roof slab finished with Tretodeck. The advantages of polymeric coatings are their exceptionally low dead

Fig. 3.8 View of a car-deck roof finished with 'Tretodeck' elastomeric coating. Courtesy: Tremco-Tretol Ltd.

weight arising from the overall thickness of less than 2 mm (approximately 2 kg/m^2), and that the material can be obtained in light colours thus reducing the thermal effect of solar radiation.

Special aggregate should be included in the final coat to provide skid resistance.

For mastic asphalt, reference should be made to Code of Practice, CP 144: Part 4 and to the technical publications of the Mastic Asphalt Council and Employers Federation.

(a) the concrete should be finished with a power float or trowel, rather than with a beam tamper, in order to provide a smooth surface;

(b) a glass-fibre tissue isolating membrane should be provided between the deck slab and the asphalt;

(c) the lower layer of mastic should be laid in two coats, each 10 mm thick, of roofing grade asphalt. On this is laid 30-35 mm of paving-grade mastic asphalt in one layer (laid as soon after the roofing grade as possible). It is advisable for added grit to be specified to increase wear and skid resistance;

(d) the disadvantages of mastic asphalt compared with suitable polymeric coatings is its black colour and hence absorption of solar radiation, and the increase in dead load resulting from the total 50 mm thickness (about 120 kg/m^2).

The relevant British Standards for mastic asphalt are:

Roofing grade: BS 6577 or BS 6925
Paving grade: BS 1446 or BS 1447

Gradients

The roof must be provided with adequate drainage outlets and gradient(s) to ensure rapid disposal of rain water. The minimum *finished* fall should be 1 in 80, and this requires the design gradient to be greater (say 1 in 40), to allow for deflexion of the deck under load, and construction inaccuracies. Similar gradients and drainage should also be provided for the intermediate decks.

Joints

All practical precautions should be taken to ensure that all joints are watertight. Construction joints should be detailed as for joints in a water-retaining structure, i.e. the recommendations in BS 8007: Code of Practice for the design of concrete structures for retaining aqueous liquids, should be followed. This means that the joints should be as monolithic as possible, and preferably provided with a sealing groove about 10 mm × 10 mm filled with an epoxy polysulphide sealant. This applies to construction joints in both floor and roof slabs. Movement joints pose special problems as they have to accommodate movement, the extent of which can only be estimated due to the many factors involved. The author's experience is that for satisfactory results, it is better to specify/use one of the few reliable proprietary systems available on the market, such as the 'Waboflex' joint marketed by Servicised Ltd, Slough, and the 'Radflex' system marketed by Radmat (London) Ltd. Figure 3.9 shows a Waboflex joint.

Needless to say, it is essential that these joints should be installed by specialist contractors recommended by the suppliers of the proprietary system selected.

Fig. 3.9 View of 'Waboflex' expansion joint in deck of multi-storey car park. Courtesy: Servicized Ltd.

Soffit surface finishes

The soffits of the decks in multi-storey car parks are normally left from the forms, i.e. it is not the intention to give any applied finish. This book is not about concrete finishes, but the author's experience is that unless both specifier and contractor know clearly what type and standard of finish is required, a contractual dispute can arise very easily. The author can only recommend that reference should be made to the following publications:

- Concrete Society, Report (1986), *Formwork, A Guide to Good Practice.*
- Property Services Agency (1988), General Specification, Work Groups A–K, Section E.20, Formwork for In-situ Concrete.
- Department of Environment (1976), Ministry of Transport, Specification for Road and Bridge Works, series 5.

Cold-store floors

The design and construction of cold stores is highly specialized and this section outlines only the basic principles involved in the design and construction of the floors of cold stores.

Cold stores vary greatly in size and in operating temperature, and consequently there is a wide range in the dead and live loads which the floor has to carry, as well as in the operating temperature. Large cold stores may be divided into a number of sections, each with their own optimum operating temperature. This necessitates the provision of an expansion/contraction joint between these different sections.

In stores of the smallest size, the moving loads will arise from foot traffic and hand-operated trolleys. In the large commercial stores the floor will have to carry heavy mechanically-operated trucks and trolleys, and high racking with resulting high point loads. Mobile racking is often used in cold stores as this increases the storage capacity compared with fixed racking.

While the majority of cold-store floors are ground-supported slabs (slabs on grade), the need sometimes arises for a cold store to be above ground level, requiring a suspended slab. With suspended floors, there is the possibility of severe condensation on the soffit of the slab, unless special precautions are taken.

A major problem with cold-store floors is the great difficulty in carrying out repairs when the need arises. As in all industrial floors, joints are particularly vulnerable to damage and sealant deterioration.

The author offers the following comments on basic factors involved in the construction of ground-supported slabs:

(a) The slab must be designed to carry the loads, moving and static, imposed on it. Recommendations are not made for the thickness of the various layers of the different materials, as these figures form part of the overall design of the floor.

(b) An adequate thickness of selected sub-base must be provided which is non-susceptible to frost; this is to eliminate the possibility of frost heave. This should be provided even though heating cables may be installed in floor system, as the possibility of breakdown of this heat control system cannot be ruled out.

(c) A separating (slip) membrane of two layers of 1000 gauge polythene sheeting, well lapped and with a breaking joint should be laid on the compacted sub-base.

(d) For a structural reinforced concrete slab, the concrete should have a characteristic compressive strength of 40 N/mm^2. If heating elements are provided they should be located in this slab, and the installation tested before the next layer (membrane and thermal insulation) is placed.

(e) On the structural slab a high-quality vapour-resistant membrane should be laid. The author believes it is justified to provide a membrane material of the highest quality, such as Heavy Duty Bituthene (Servicised Ltd), or Hyload 75 (Ruberoid, Enfield), which are sheeting materials. If an in-situ brush-applied membrane is preferred, then this can be bitumen, coal tar or polymer resin based, applied in two coats.

(f) The thermal insulation should be laid on the membrane. This should be able to resist the floor loading without deformation or creep, must be resistant to moisture and must retain its thermal insulating properties at low temperatures. A suitable material would be a selected grade of Styrofoam (extruded polystyrene foam boards). All joints should be properly taped and the boards laid as directed by the suppliers.

(g) The top concrete slab, which forms the wearing surface of the floor, is laid on the thermal insulation. There are differences of opinion as to whether this concrete should be air entrained. Those who feel that air entrainment is not necessary, claim that commercial cold stores operate indefinitely at a consistent low temperature (say −25 °C), and so freeze-thaw cycles do not occur. Also it is better to specify a high characteristic strength of say 50 N/mm^2, which would require a cement content well above 400 kg/m^3, and this would create considerable problems with the dispersal in the mix of an air-entraining agent. Joints in this slab should be kept to a minimum as they are vulnerable to damage. It is advantageous if they can be located under fixed racking.

(h) The floor should be finished to line, level and 'flatness' to suit the operating traffic.

Figure 3.10 is a sketch showing the construction of a cold-store floor.

As the cold store will operate for very many years at a very low temperature (–25°C to –30°C), it is reasonable to assume that the concrete slab which forms the wearing surface will have a temperature throughout its depth of well below freezing. In other words, the concrete will be virtually in a state of permafrost. If flexible sealants have been used in the slab joints, they will not remain flexible at this temperature.

Shrinkage-compensating cements are available in the US, and have been used successfully in concrete for the floors of cold stores. A reference is included in the Bibliography at the end of this Chapter.

Some information is given in Chapter 7 on the principles of repairs to concrete floors in cold stores, in which there are inherent difficulties.

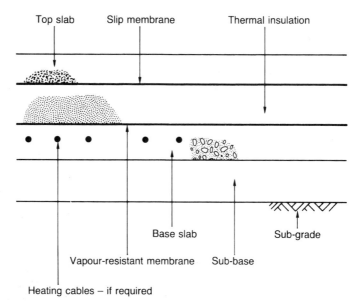

Fig. 3.10 Sketch showing the construction of a cold-store floor.

Bibliography

Publications and papers detailed in the text are not included here.

American Concrete Pavement Association (1989) *Fast Track Concrete Pavements*, Technical Bulletin TB-004.0 IT, ACPA, Arlington Heights, Illinois, USA

BS 8110 Code of Practice for the structural use of concrete, British Standards Institution, Milton Keynes.

BS 5328: 1990 Concrete: Parts 1–4, British Standards Institution, Milton Keynes

DD 83 Assessment of the composition of fresh concrete: draft for development, British Standards Institution, Milton Keynes

BS 6399: Part 1 Design loadings for buildings; Code of Practice for dead and imposed loads, British Standards Institution, Milton Keynes

BS 648 Schedule of weights of building materials, British Standards Institution, Milton Keynes

BS 8204 In-situ floorings: Parts 1 and 2 British Standards Institution, Milton Keynes

Building Research Establishment (1983) *The Incidence of Accidental Loadings in Buildings* IP8/83, p. 4

Building Research Establishment (1971), *Floor Loading in Retail Premises – The Results of a Survey*, CP25/71, p. 37

Building Research Establishment (1987) *Classes of Imposed Floor Loads for Warehouses*, IP 19/87, p. 4

Building Research Establishment (1960) *Floor Finishes for Factories*, HMSO, London, p. 11

Building Research Establishment (1969) *Floor Screeds*, Digest 104, p. 8

Building Research Establishment (1979) *Estimates of Thermal and Moisture Movements and Stresses: Parts 1, 2 and 3*, Digests 227, 228 and 229, July to Sept. 1979

Cement and Concrete Association (1960) *Specification for Granolithic Floor Toppings Laid on In-situ Concrete*, p. 21

Chaplin, R.G. (1986) *Mixing and Testing Cement-Sand Floor Screeds*, Cement & Concrete Association, p. 28

Harrison, R. and Malkin, F. (1983) On-site testing of shoe and floor combinations. *Ergonomics*, **26**(1), 101–108

James, D.I. (1984) *Slip Resistance Tests for Flooring, Two Methods Compared*, Rubber & Plastics Research Association (RAPRA), Members Report No. 94, p. 16

Lea, F.M. (1970) *The Chemistry of Cement and Concrete*, 3rd edn, Edward Arnold, London, pp. 542–546

National Federation of Terrazzo, Marble and Mosaic Specialists, Handbook, 1992

Portland Cement Association (USA) (1987) *Concrete Information – Concrete Slabs Surface Defects: Causes, Prevention Repair*, p. 5

Proctor, T.D. and Coleman, V. (1988) Slipping, tripping and falling accidents in Great Britain – present and future. *Journal of Occupational Accidents*, **9**, pp. 269–285.

Roberts, J.J. (1973) *The Crazing of Concrete*, Cement & Concrete Association Technical Report 42.480, May, p. 19

Roberts, J.J. (1975) The crazing of concrete, Paper at Inter-Association Colloquium, Liege, p. 7

Williams, A. and Clements, S.W. (1980) *Thermal movements in the upper floor of a multi-storey car park*. Cement & Concrete Association Technical Report 539, p. 23.

Chapter 4
Finishes for concrete floor slabs: tiles, slabs and sheeting

Introduction

The materials dealt with in this chapter are:

- *tiles*: terrazzo, ceramic, marble, marble conglomerate, composition block, ceramic pavers
- *sheeting*: PVC, rubber, linoleum, cork
- *timber flooring* (on concrete base slab)

The reason for including such materials as vinyl, linoleum, rubber and cork, is that special precautions have to be taken when these materials are laid on concrete and cement-based screeds due mainly to moisture in the concrete/screed. Also, these materials offer appreciably less impact resistance than rigid tiles and slabs, and require the substrate to be stronger and have higher impact resistance.

Timber flooring is resistant to impact but is vulnerable to rising damp and requires special precautions. See Fig. 3.1 for an illustration of the terms used.

Terrazzo tiles

General considerations

The relevant Code of Practice is BS 5385: Part 5: The design and installation of terrazzo tile and slab, natural stone and composition block flooring. It replaces the relevant parts of CP 202. There is also a handbook issued by the National Federation of Terrazzo, Marble and Mosaic Specialists.

The British Standard for terrazzo tiles is BS 4131. Terrazzo tiles made and laid in accordance with the relevant British Standards are very hard wearing and are produced in a wide range of attractive colours (Fig. 4.1). The tiles are not supplied 'fully finished' and are

Fig. 4.1 A view of terrazzo tiles in a large shopping centre. Courtesy: A. Quiligotti PLC, Stockport

ground, regrouted and polished not less than 3 days after the completion of the original grouting of the joints.

Section 3 of BS 5385 Part 5 covers design, which the author interprets to mean that where there is a specification for the work, this document should include the relevant clauses in this section. In other words, the inclusion of the recommendations set out in Section 3 into the Contract is the responsibility of the designer/specifier. This would not apply in cases where the contractor was responsible for both design and construction.

Materials and construction

General
The tiles should comply with BS 4131: Terrazzo tiles, and reference should be made to that Standard for the detailed requirements. The tolerances given on the dimensions of the tiles should be noted, particularly the ± 3 mm on the nominal thickness because unless this is allowed for in deciding on the thickness of the bedding when an

adhesive is used, unfortunate results can arise. This is referred to below under 'Tolerances'.

The author's experience is that the recommendations in clauses 17 to 19 relating to independent testing are not always carried out, even on large projects. Once the tiles are laid the recommended sampling procedure cannot be implemented.

The Code of Practice, BS 5383: Part 5 gives detailed recommendations for the laying of terrazzo-tiled floors and the author does not intend to repeat them here, but only to draw attention to those matters which in his experience are likely to result in complaints about the final result.

BS 4131 requires that the thickness of the top (wearing) layer of the tiles should be at least 6 mm after grinding, and consist of graded marble and Portland cement. The tiles should be at least 28 days old when they are laid, to allow for a significant amount of shrinkage to take place before laying. The joints between the tiles are made with a cement-based grout which also shrinks.

The time schedules recommended in clause 3.3 of the Code should be adhered to. When the contract is behind schedule, as is frequently the case, these recommendations tend to be ignored. When this occurs it is in the tile layer's interest to ensure that this is clearly recorded in writing.

Grouted joints

The joints between the tiles are narrow (about 2-3 mm) and so the floor finish as a whole, should be considered rigid (non-flexible). This fact is of great importance when movement is likely to occur in the structural floor slab. Unless this is allowed for in the tiling, the result is likely to be opening of joints and loss of grout filling and/or cracking of the tiles. Where the tiles are used in areas such as airport terminals, shopping centres, etc., they are subjected to cleaning by mechanical cleaners using rotating brushes, water containing a cleaning liquid and a suction device to remove surplus water. As the joints are fine (2–3 mm), the Code recommends a neat cement grout; this will shrink and crack and can be dislodged by the cleaning machines. The fine transverse cracks thus formed become filled with dirt and this can seriously disfigure an otherwise well laid and attractive floor. To overcome this problem, the author recommends the use of proprietary 'non-shrink' grouts. The use of these grouts has proved to be successful. The formation of these shrinkage cracks when neat cement grout is used can lead to serious disputes between the specialist sub-contractor, the main contractor, the architect and the employer.

Movement joints

All buildings move to a greater or lesser degree. This overall movement is made up in the main of foundation movement, initial and cyclic thermal movement, creep under load, deflexion under load and drying shrinkage. The final overall movement cannot be predicted accurately. The movement of a ground floor slab (slab on grade) will be different to that of a suspended slab.

The recommendations in the Code for the location and detailing of movement joints are practical and should be followed. Briefly summarized, movement joints should be provided:

(a) to coincide with movement joints in the structure;
(b) at the perimeter of the tiled area, and where the tiling abuts on units which offer restraint to movement, such as columns, kerbs and plant fixed to the base slab;
(c) over supporting walls and beams;
(d) as intermediate joints (which are really stress-relief joints) at 8–10 m intervals in both directions. The author's experience suggests that this spacing is rather close, suggests 10–12 m. In addition, there should be flexible movement joints at about 25–30 m centres in both directions.

The joints should be carefully detailed to accommodate movement and to protect the edges of the tiles and support the side of the tile bedding and screed. Proprietary jointing systems are to be preferred to 'in-house' designs (see examples in Fig. 2 in BS 5385: Part 5). Stainless steel cover strips (fixed on one side of the joint) can be a useful solution to the problem of sealant protection when using a non-proprietary jointing system (see Fig. 4.2).

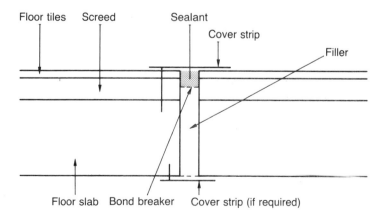

Fig. 4.2 A sketch showing a cover strip to protect the sealant in a movement/expansion joint in a tiled floor. (Side support to the screed is not shown)

Separating layer

Associated with the problem of movement in the structural slab and tiling, is the need or otherwise to provide a separating layer, either between the base slab and the screed, or between the screed and the tile bed (if a screed is provided), or between the base slab and the tile bed if a screed is not provided.

The author's experience is that a separating layer should only be provided when it is really necessary, i.e. when significant movement is anticipated between the structural slab and the tile bed. In ground slabs the separating layer should not be considered as a substitute for a proper damp-proof membrane. However, the dpm may be considered as an effective separating layer, provided it is constructed so as to allow the upper layer (screed or tile bed) to slide on it.

The Code, at clause 34.2.1, claims that when a 'semi-dry' mix is used for bedding the tiles, there is a very weak bond between this type of bedding mortar and the substrate on which it is laid. This may be correct, but the author considers that a mix with this characteristic is unlikely to possess high impact resistance, and therefore should not be used for heavily-loaded floors. Semi-dry mixes are very difficult to compact, and proper compaction of the mortar is essential to resist impact damage.

When a polyethylene sheet is used as a separating layer, it is important to ensure that the surface of the concrete is reasonably flat and free of ridges, otherwise the tile bed and tiles may 'rock', resulting in opening of joints and cracking of tiles.

Flooring beds

The Code recommends that suitable beds for terrazzo tiles are cement/sand mortar and the 'semi-dry' mix.

Table 4 on page 23 of the Code provides recommendations for the type and thickness of the beds for flooring materials covered by the Code. This Table indicates that a thick bed adhesive can be used, provided the manufacturer confirms its suitability. It should be noted that the minimum thickness of a cement/sand mortar bed is 15 mm, with a maximum of 25 mm, while for a 'semi-dry' mix, the minimum is 40 mm and the maximum is 70 mm. These figures are very important when dealing with finished floor levels and the levels of the base concrete. It is very rare that an adhesive is used for terrazzo tiles.

The anticipated floor loading must also be taken into account when considering the thickness of the tiles and type and thickness of the tile bed.

Section 23 of the Code deals with loading in some detail. However, the author feels that the division of loading into only two categories – light and heavy – is an unfortunate over-simplification, and he cannot agree that pedestrian traffic in 'shopping malls' should be considered as 'light loading'. Such floors have to carry literally hundreds of thousands of pedestrians every year, as well as heavy cleaning equipment. It would be prudent to consider shopping centres, department stores, supermarkets and similar areas, as heavily-loaded floors.

For heavily-loaded floors (as defined above), it is most important that the tile bed, based on cement/sand mortar, should be thoroughly compacted, and for this reason the author does not recommend the use of a semi-dry mix, as such mixes are very difficult to compact (see Fig. 4.3).

Fig. 4.3 Laying terrazzo tiles. Courtesy: A. Quiligotti PLC.

When the tiles are bedded on a screed, the impact resistance of the screed should first be checked by means of the BRE Screed Impact Tester (see relevant sections in Chapters 3 and 6). Subject to practical interpretation of the test results, the author recommends a maximum indentation of 3 mm for floors subjected to heavy wear.

Clause 22.3 of the Code refers readers to Appendix C of Part 3 of BS 5385 for recommendations for design and laying of screeds.

However, when the tiles are bedded on a cement/sand mortar it is also important that for heavily-loaded floors the density (degree of compaction) of the bed should also be checked. Such floors are usually large in area and laboratory tests and site trials should be carried out to determine a reasonable bulk density that can be achieved under site conditions. This is likely to be in the range of 1900-2150 kg/m^3. Density checks on site can be made by hammering steel cylinders of known volume and weight into the plastic mortar as soon as compaction is completed. A simple calculation will disclose the bulk density of the mortar.

This additional site control may appear exaggerated, but it should be evaluated with the cost, inconvenience and delay resulting from failure to pass the specified BRE test.

A further point to help ensure good-quality mortar is that the mortar should be weigh batched. This can be arranged by using the sand prepacked in 25 kg or 50 kg bags. For small jobs careful mixing by hand can be accepted, but for larger projects the mortar should be mixed in a paddle type or forced action mixer. A free-fall concrete mixer should not be used. The mix proportions for the bed should be in the range of 1:3.5 to 1:4.5 by weight (approximately 1:3 to 1:4 by volume).

Where the bed is laid on the concrete base slab and good bond is required, the surface of the concrete should be clean, damp and preferably the coarse aggregate should be slightly exposed. Also, a bonding agent can be used to advantage.

Tolerances

Tolerances on dimensions must always be taken into account, but unfortunately this is not always done at the design/specification stage. Recommended tolerances on the finished floor are given in Clause 33.4 of the Code. It is specially remarked in the Code that: 'The accuracy of the surface may be limited by the dimensional tolerances of the flooring units.' This refers to the tolerances on the dimensions of the terrazzo tiles, which has been referred to earlier in this section. There is also the permitted tolerance on the surface of the substrate (see the relevant section in Chapter 3). Thus the tolerances on the substrate and on the tile thickness can be cumulative and this must be given proper consideration in fixing and checking finished floor levels.

It is necessary to remember that:

(a) the closer the tolerances specified, the more expensive the work;
(b) the specified tolerances should be appropriate for the use to which the floor will be put;
(c) the method of checking the tolerances actually achieved on site should be detailed clearly – different methods will usually give different results.

Over large floor areas a tolerance of ± 15 mm is allowed. Local variations in level are restricted to ± 3 mm under a 2 m straight edge, but the method of carrying out the test is not detailed as it is in Part 3 of the Code for Ceramic Floor Tiles.

Lipping between tiles is limited to 1 mm for joints less than 6 mm wide, and to 2 mm for joints 6 mm or more in width. As the joints between terrazzo tiles are normally a maximum of 3 mm in width, the maximum 'lip' between adjacent tiles is 1 mm. This should be readily achieved with terrazzo tiles as they are ground and polished on completion. However, problems can arise if the tiles are bedded on 6 mm thick-bed adhesive, as this provides insufficient depth to allow for irregularities in the base slab.

Slip resistance

An ideal floor surface should be resistant to slipping (i.e. have a high coefficient of friction) and at the same time be virtually self-cleansing when washed. The author does not know of any such flooring material. This means that a compromise has to be reached between the two extremes, high slip resistance and ease of cleaning. This rather simple technical fact is often not appreciated by specifiers and building owners.

The actual slip resistance depends on the surface characteristics of the two materials in contact, namely, the floor surface and the soles of the footwear worn by the persons using the floor. A considerable amount of research has been carried out on the slip resistance of floor surfaces and materials used for footwear. The following organizations have special experience and knowledge in this field:

British Ceramic Research Limited (BCRL)
Rubber and Plastics Research Association (RAPRA)
Shoe and Allied Trades Research Association (SATRA)

BCRL have developed equipment for measuring the coefficient of friction of any type of floor surface *in-situ*. This machine is known as the TORTUS and can be purchased or hired. The Transport and Road Research Laboratory (TRRL) have developed equipment for measuring the skid resistance of surface of roads and external paving, and this equipment can be used for large floor areas (see Fig. 2.13).

The SATRA laboratory test is considered by some authorities to provide the closest representation to actual on-site slipping. However, it is a laboratory test and the author feels it is most important that test methods should be suitable for site application when the assessment of the actual slip resistance of an industrial or commercial floor is required.

The finishing process
The traditional method of finishing terrazzo tiles is:

(i)   grinding (twice, using first coarse and then finer grinding stones);
(ii) washing and regrouting, followed by polishing with fine grit stones;
(iii) usually, but not invariably, the application of a surface sealer.

The final result is a very hard wearing attractive floor, but one liable to be very slippery when wet.

Floor gradients

The majority of floors are intended to be laid level (see the section on 'Tolerances' earlier in this chapter). However, it is normal for industrial floors where 'wet' trades are used, to be laid to gradients. The terms, 'level', 'flat' and 'gradient' are not always understood and this can, and does, give rise to construction disputes. The insertion of floor channels and gulleys is essential in such circumstances, with the floor laid to adequate gradients (falls) to the channels. The selected gradient must be sufficient to provide proper drainage of the floor surface, but at the same time must not be sufficiently steep as to be dangerous to the persons using the floor. Ponding is always considered undesirable, but in practice it is virtually impossible to eliminate it completely. It is the author's experience that a gradient of 1 in 80 is adequate for the actual finished gradient; the design gradient is greater. With a reasonably non-slip surface this is unlikely to present any hazard to the users. It must also be kept in mind that an increase in slip resistance will be accompanied by an increase in difficulty in cleaning the surface. These basic facts are of great importance in the food industry, and a sensible compromise has to be reached; it is desirable for the building owner to be involved with this decision making. Any floor surface that becomes contaminated with grease or oil or very fine dry powder will be slippery and will possess a low coefficient of friction.

Ceramic tiles and mosaics

General

The relevant Code of Practice is BS 5385: Part 3: The design and installation of ceramic floor tiles and mosaics. It replaces relevant sections in CP 202. Recommendations for mosaic floors have been added to the Code. The comments made by the author on Section 3 of the Code BS 5385: Part 5: Terrazzo tiles, also apply here.

Ceramic tiles are covered by BS 6431: Part 1, and are in two main classes: extruded tiles and dust-pressed tiles. They are further divided according to their water absorption and whether they are glazed, partly glazed or unglazed. Generally, the pressed tiles are made to finer dimensional tolerances and therefore the joints can be narrower than required for extruded tiles (see Fig. 4.4).

Grouted joints between tiles

Clause 23.5 of the Code recommends that joints should not be less than 3 mm wide; generally they are between 3 and 6 mm wide. Joints are grouted with a cement/sand grout. The grading of the sand

Fig. 4.4 View of a high-quality ceramic tiled floor. Courtesy: British Ceramic Research Ltd.

depends on the width of the joints. (Section 9 of the Code); a coarser sand being used for the mortar in the wider joints. With good-quality tiles which are properly laid, joints are the most vulnerable part of the floor, and this applies particularly to joints which exceed about 6 mm in width. Considerable damage can occur to wide joints from stiletto heels. The load from an average stiletto heel measuring 7 mm diameter is likely to be about 1.25 kg per mm^2, without any allowance for the effect of impact.

The author therefore recommends that all joints in tiled floors which have to carry heavy pedestrian traffic should be grouted with low shrinkage, high impact-resistant proprietary grout.

Movement joints

This subject is covered in considerable detail in Section 19 of the Code to which readers should refer, particularly the diagrams showing suggested joint details.

The Code recommendations are essentially similar to those for terrazzo tiles in BS 5385: Part 5: Section 37, and which have been summarized earlier in the chapter.

Methods of laying the tiles

The definitions are illustrated in Figure 3.1 (in BS 5385: Part 3). The base on which the tile bed is laid is generally a screed or the concrete base slab itself.

The time schedules referred to under 'Terrazzo tiles' earlier in this chapter should be particularly noted and adhered to as the recommendations apply to ceramic tiles; the author's comment on the implementation of the schedules should also be noted.

If it is intended to lay the tile bed directly onto the concrete base slab, the type of bedding material and the standard of accuracy (tolerances) of the finished concrete have to be taken into account in order to obtain the required tolerances on the finished surface of the tiles. Also, consideration must be given to the loading conditions, i.e. the type and magnitude of the traffic. Section 15 of the Code discusses this subject in some detail and is a great improvement on the previous Code (CP 202). The recommendations in BS 5385: Part 5: Terrazzo tiles, Section 23, are essentially the same, and the author's comments are applicable here, i.e. that floors of shopping centres, department stores and similar, should be considered as heavily loaded, and the tiles and tile bed specified accordingly.

Tolerances

Tolerances are covered by the recommendations in Clause 23.4 of the Code. These recommendations deal with differences in level over the whole of the floor, and allow a tolerance of ± 15 mm, except at partitions, door openings and where machinery has to be fixed to the floor. Local variations are measured by means of a 2 m long straight-edge with 3 mm thick feet at each end. The straight-edge should not be obstructed by the tiles and no gap should be greater than 6 mm, i.e. a tolerance of ± 3 mm. It should be particularly noted that the method used is different to that for screeds as described in Appendix D of the Code.

Lipping at joints is restricted to 1 mm for joints less than 6 mm wide, and to 2 mm for joints 6 mm or more in width.

Floor gradients

The comments and recommendations regarding floor gradients given above under 'Terrazzo tiles' are also applicable here, but it must be kept in mind that ceramic tiles are a different material to terrazzo. Unglazed ceramic tiles are slip resistant. Glazed tiles that have not been specially manufactured to be slip resistant are very slippery when wet.

Any floor surface that becomes contaminated with grease or oil or a very fine dry powder is likely to be slippery and to possess a very low coefficient of friction.

Marble tiles and slabs

Marble tiles and slabs are included in Section 6 of BS 5385: Part 5.

There is no British Standard for marble tiles and so there are no authoritative recommendations for dimensional tolerances and this should be discussed with the suppliers. Some detailed comments on the practical aspects of dimensional tolerances with particular reference to the implications of tolerances on thickness, are given in the section on terrazzo tiles.

The basic principles in laying terrazzo tiles and ceramic tiles also apply to marble. Movement joints, location and detailing are similar, with the exception of small, and in the author's opinion rather unimportant, changes in the detail for perimeter joints. As far as bedding is concerned, the Code recommends a cement/sand mortar; provided that the marble units do not exceed 15 mm in thickness, a thick or thin bed adhesive may be used. The advice of the adhesive manufacturer should be sought before specifying a particular type and/or laying thickness. The thickness of the marble tiles and the thickness of the adhesive are important as this will be fundamental in determining the finished floor level and surface tolerances.

When the units are laid on a cement/sand mortar bed, the backs of the units should be coated with a white Portland cement slurry which can, with advantage, incorporate a bonding agent. The units are laid with narrow joints up to 2 mm wide, and carefully tapped into position.

Grouting is normally with neat cement grout as the joints are very narrow (1–2 mm) and is carried out in a similar way to grouting terrazzo tiles. For reasons already given under 'Terrazzo tiles', the author strongly recommends the use of proprietary 'non-shrink' grouts.

Marble units are laid as 'fully finished' and are normally not ground and polished on completion. Cleaning may be carried out by machine using water and a neutral detergent, but the recommendations of the suppliers should be obtained and followed, and these usually include periodic treatment with a proprietary sealer or wax. The use of soap is not recommended.

Marble conglomerate tiles (reconstituted marble)

There is no British Standard for marble conglomerate tiles and the Code, BS 5385: Wall and floor tiling, does not refer to it.

The tiles are made of pieces of marble unsuitable for cutting into slabs and tiles. The pieces of marble are fed into large moulds and the voids are then filled in with marble chippings and fine marble as a filler. The whole is then impregnated under pressure with a resin, usually a polyester. After curing, the moulds are removed and the hardened block is sawn into slabs or tiles, which are normally honed or polished. The tiles and slabs are therefore received on site as 'fully finished'.

The sizes of the tiles vary, but are usually in the range of 300 × 300 × 9 mm to 600 × 300 × 14 mm.

As there is no National Standard for these tiles, there are no authoritative recommendations for dimensional tolerances. This is important in deciding on the allowance for depth of tile bed; the author recommends that the tolerances given in BS 4131 for terrazzo tiles, i.e. ± 3 mm, should be allowed for marble conglomerate tiles and slab. Some detailed comments on tolerances on the floor surface are given in the sections on terrazzo tiles and ceramic tiles.

Due to the variation in the colours and sizes of the pieces of marble in the units, the appearance is very attractive. Unfortunately, the author's experience is that under heavy pedestrian use, the floor tiles are not durable. The main defect is the loss of small pieces of marble from the wearing surface. These small depressions become dirt stained and the overall appearance of the floor deteriorates.

Marble itself is a very variable material and due to the method of manufacture of these tiles, this variability is increased considerably.

The author has heard it suggested that in these tiles, the lighter coloured pieces of marble are less hard wearing than the darker marbles, but he has found no support for this from investigations of large commercial floor areas. Suppliers usually recommend that the tiles should be laid in a similar way to marble tiles, but special care should be taken in laying to ensure solid and uniform bedding. Joints between tiles are narrow (1–2 mm) and grouted with neat cement grout. For the reasons stated in the section on terrazzo tiles, the author strongly recommends the use of proprietary 'non-shrink' grouts.

Movement joints should be incorporated in the floor, as recommended for marble tiles.

Maintenance is particularly important due to the tendency mentioned above for pits to form in the floor surface. The use of mechanical cleaning/washing machines should be avoided whenever possible as they tend to aggravate the 'pit-forming' characteristic of these tiles. Careful use of water with a neutral detergent followed by regular sealing or waxing is recommended. The use of soap is not recommended.

Ceramic pavers

The best known ceramic pavers for chemically-resistant floors are the 'Nori' pavers and tiles made by the Accrington Brick and Tile Co, Accrington (Fig. 4.5 (a)). These pavers and tiles are made from silica alumina shale, which is extruded or moulded to the sizes and shapes required, and fired at a temperature of about 1035°C. The standard brick is 220 × 105 × 73 mm; the plain tiles have a range of dimensions, including 305 × 152 × 50 mm and 220 × 105 × 38 mm. Special sizes and shapes can be supplied to order. The relevant British Standard is BS 3921: Specification for clay bricks. Reference can also be made US Standard, ASTM C410.

Where chemical resistance of the floor is not required, the units can be bedded and jointed in a cement/sand mortar or 1:3 to 1:4 by volume. The workability and water resistance of the mortar will be improved by the addition of a styrene butadiene emulsion (10 litres emulsion to 50 kg cement).

For medium chemical resistance, the units should be bedded on a cement/sand/SBR mortar, with the joints completely filled with an epoxy resin mortar.

For maximum chemical resistance, the units should be fully bedded onto a high-quality chemically-resistant membrane using a high chemically-resistant mortar such as epoxy, furane, polyester, etc., and the joints carefully filled with the same mortar.

The joints between the pavers are narrow (3 mm) and the bedding mortar is about 5 mm thick.

Movement joints should be provided to cater for cyclic thermal movement. While the exact location of movement joints depends on the floor layout, they are normally required at the perimeter of the floor, around projections through the floor such as columns, plinths for machinery bases, manholes, etc., and in line with movement joints in the base slab. For large areas, movement joints are also provided at about 6 m centres in both directions. Special care is needed in the selection of the sealants for these joints. A joint width of about 10 mm is usually adequate except for the structural movement joints in the building.

Flexible sheeting and tiling

Introduction

The materials covered in this section are: flexible PVC sheet and tiles; backed flexible PVC sheet and tiles; thermoplastics and semi-flexible PVC tiles; linoleum sheet and tiles and cork carpet; cork tiles; and

Fig. 4.5(a) View of a floor in a chemical plant finished with Nori acid-resisting pavers. Courtesy: Armitage Brick Ltd.

Fig. 4.5(b) A warehouse floor finished with Altro 'Hyload' quartz-reinforced vinyl tiles. Courtesy: Altro Floors. Letchworth

rubber sheet and tiles, when these materials are laid on concrete or cement-based screeds. The laying of these materials on timber or other substrates is not dealt with in this book.

The relevant British Standard is BS 8203: Code of Practice for the installation of sheet and tile flooring. The code gives useful basic information on the following properties of these flooring materials:

- wear resistance and indentation
- resistance to moisture
- slip resistance
- flexibility
- thermal properties (user comfort)
- impact sound transmission
- staining and colour fastness

Readers are referred to the Code and to the literature issued by the manufacturers for more detailed information on each type of product.

The responsibility for the physical failure of flexible types of flooring is often placed on defects/shortcomings in the substrate on which they are laid.

General recommendations for the laying of concrete floor slabs and cement-based screeds have already been given in Chapter 3, but the author's experience suggests that *additional* discussion and recommendations would be useful when relatively thin sheet and tiles are laid on concrete and screeds. Special problems arise from moisture in the cement-based substrate. The main source of this moisture is the mixing water, part of which slowly rises to the surface and evaporates; added to this inherent moisture is moisture rising from the sub-soil in slabs laid on the ground when a damp-proof membrane has not been provided in the floor construction.

Floors in contact with the ground

When the floor slab is in contact with the ground, the provision of a high-quality damp-proof membrane, properly installed, is essential. The Building Regulations 1991 require the provision of such a membrane in almost all cases where a Building Permit is required. However, a Building Permit is not required for the provision of new flooring material on an existing floor slab and the need to provide a dpm is then often overlooked. When any of the above materials are installed in a floor slab in contact with the ground where a dpm is not provided, serious trouble is almost inevitable. The flooring material, together with the adhesive used, will form an effective barrier to the evaporation of moisture from the substrate, resulting in the accumulation of water under the floor covering, which in turn gives rise to debonding, blistering and general deterioration.

The Code (BS 8203: Section 8) stresses that the adhesives used cannot be considered effective as a dmp. It is generally worthwhile to consider the use of an adhesive that is not moisture sensitive when applied to ground floor slabs/screeds, even though a dpm is provided.

Ground floors and suspended floors

The suppliers of the materials dealt with in this section invariably require that the moisture content of the substrate on which the sheeting/tiles are to be laid should be measured and must not exceed a prescribed figure. This problem is discussed at some length in Appendix A of BS 8203. The method described is applicable to both ground floor slabs and suspended slabs. The determination of the moisture content of the concrete/screed involves measuring the relative humidity of a small volume of air entrapped between the surface of the substrate and the measuring apparatus, namely a hygrometer. When the relative humidity of this entrapped air is 75% or less, it is generally accepted that the substrate is 'dry' and the sheeting/tiles can be laid on it. The figure of 75% is a good general guide, but the recommendations of the material supplier should be included in the specification, together with instructions for the method of measurement.

A quick, but rather less accurate way of obtaining an approximate measure of the moisture content of cement-based screeds and concrete is to measure the electrical conductivity. A suitable piece of equipment measures the conductivity between two electrodes which have been inserted into two predrilled holes, 25 mm deep, in the substrate. The presence of chlorides and some admixtures in the screed/concrete can adversely affect the accuracy of the readings. When doubt is expressed about the reliability of the test results, a check should be made by either using the standard hygrometer or by covering the area of test with polythene sheets (held down around the perimeter) for 24 h and then drilling new holes and measuring the conductivity. If the new readings are significantly higher than the original ones, then this would indicate that the first set of readings should be disregarded as the substrate had not dried out sufficiently.

It should be noted that it is not the actual moisture content of the substrate that is important, but the water vapour leaving the top surface of the substrate. Water that is locked into very small pores is likely to produce low vapour pressure, which is unlikely to harm flooring laid on the surface. The author understands this is the 'secret' of the proprietary screeds which are claimed to possess very rapid drying characteristics.

This brief discussion on problems arising from the moisture contents in screeds and base slabs would not be complete without reference to 'osmosis' as a cause of blistering of in-situ resin-based flooring and some sheet floorings. Briefly, the blisters are found to be filled with a liquid in a state of compression. When such blisters are punctured, a jet of liquid may spurt several metres into the air. A detailed report on this phenomenon is contained in a paper by W.J. Warlow and P.W. Pye, in the *Magazine of Concrete Research*, September 1978.

Surface regularity (smoothness) of the substrate and variation in the thickness of screeds

Three other items need special attention, and these apply to ground floors and suspended floors. These items are the standard of surface regularity of the substrate on which the flooring will be laid; the tolerances on the variation in surface levels of the base slab; and the variation in the thickness of the screed when there is one. A reasonable required standard for surface regularity and the specified tolerance of thickness of the levelling screed are both closely related to the thickness of the floor covering. This can vary from 2 mm for flexible PVC sheeting to 8 mm for heavy-duty cork tiles.

Flexible PVC sheeting 2–3 mm thick is used extensively in hospitals where impact and wear is high, but the material itself provides little protection to the substrate against impact. It is therefore obvious that the impact resistance of the screed is of primary importance if long-term durability is to be ensured. To help ensure long-term resistance to impact, it is necessary to carry out a systematic test with the BRE Screed Tester; for detailed information on testing and interpretation of results, reference should be made to BRE Information Paper IP11/84. The author recommends that for screeds that will receive thin sheet and tiles, indentations should not exceed 4 mm; for heavily-loaded floors such as hospital corridors and the like, this figure should be reduced to 3 mm. For these permitted indentations, the recommendations in IP11/84 on interpretation of test results, should be followed.

The recommended 'nominal' thickness of cement/sand floor screeds laid on and bonded to a hardened concrete base slab is 40 mm; taking into account normal tolerances on a power floated base slab, this thickness of 40 mm should help ensure that the minimum thickness of the screed is not less than about 25 mm.

The recommended 'nominal' thickness of many proprietary screeds is appreciably less than 40 mm and then tolerances on surface regularity of base slab and screed become very important.

The likely impact resistance of screeds of the same type is related to the thickness, and this should be kept in mind.

Rippling of thin flexible sheet and tile flooring

Rippling is a disconcerting defect as it spoils the appearance of an otherwise attractive floor. The ripples form in the flooring along the line of cracks in the screed and/or bay joints, in those floors which incorporate a dpm between the base slab and the screed. The defect appears in the very early life of the floor, namely within one to four weeks after the flooring is laid. A detailed investigation by the Building Research Establishment was reported in BRE Current Paper CP94/74 to which reference should be made for complete details. The conclusions are summarized here. The upper part of the screed loses moisture much more rapidly than the lower part and this results in the curling of the edges of the screed along the line of bay joints and cracks. The presence of the dpm effectively prevents bond between the substrate and the screed. When the flexible flooring is laid on the screed the rate of loss of moisture from the surface is substantially reduced and so, as the moisture in the lower part of the screed slowly moves upwards, the moisture content of the upper part of the screed is increased and reverses the drying shrinkage which had previously occurred. The very slight bulge in the flooring is in many cases not seen unless specially searched for. The increase in moisture content in the upper part of the screed causes it to expand and the drying shrinkage curling is eliminated and a ripple is formed in the flexible flooring.

Preventive measures are difficult because the ripples only form under the conditions briefly described above, and certainly not at every crack and bay joint. The author considers the most practical approach is to control the thickness of screeds laid on dpms to the minimum recommended (50 mm) and, if possible, allow sufficient time for the full depth of screed to dry out before the flooring is laid on it.

Brief information on cork, linoleum, rubber and flexible PVC sheet and tiles

Cork tiles

Cork tiles are cut from compressed granular cork blocks. They are resilient and 'warm' and will stand up to reasonably hard wear, but are not recommended for very severe wear conditions. When laid on a ground floor slab in contact with the ground, a dpm is essential. It is generally recommended that the tiles should be laid on an adhesive and secured with hardened steel pins.

The density and thickness are directly related to the durability; for heavy wear, a minimum thickness of 8 mm and a minimum density of 500 kg/m^3 is recommended.

The relevant British Standard is BS 6826: Linoleum and cork carpet sheet and tiles,

Linoleum sheet and tiles

Linoleum has been on the market for a very long time and has proved to be a very 'comfortable' and popular flooring material in sheet and tile form. It is made from linseed oil, resins, wood flour, fillers, cork and pigments, and is pressed onto a canvas or glass-fibre backing. It varies in thickness from about 1 mm to 4.5 mm; a minimum of 3 mm is recommended to resist heavy wear.

A dpm is essential when laid on concrete slabs in contact with the ground. It is more durable when it is bonded to the sub-floor with a suitable adhesive. The joints between the sheets can be hot or cold welded.

The relevant British Standard is BS 6826: Linoleum and cork carpet sheet and tiles.

Rubber sheet and tiles

There is no British Standard for rubber sheet and tiles for flooring. The tiles and sheets are hard wearing, reasonably resistant to many dilute acids and alkalies, anti-static, and dimensionally stable. Synthetic rubbers of Neoprene or EPDM are normally used. Natural rubber is adversely affected by petroleum oils, grease and fats.

Both sheet and tiles are best fixed with an adhesive recommended by the suppliers (an epoxy-based adhesive is often used). The thickness is in the range 2–6 mm.

Flexible PVC sheet and tiles

The relevant British Standard is BS 3261: Unbacked flexible PVC flooring, which covers both sheet and tiles.

The PVC resin is mixed with stabilizers, pigments and fillers, and the sheets are 1.5 mm thick and upwards. The joints can be welded with PVC rods. It is a very popular type of flooring and is used in hospitals, department stores, etc., as it is resistant to oil, grease and a wide range of chemicals, but is easily damaged by cigarette burns, and cut by sharp objects.

An informative and useful paper was written in 1967 by W.J. Warlow, F.C. Harper and P.W. Pye of the Building Research Establishment as Miscellaneous Paper No. 13 on the resistance to wear of flooring materials. It deals with sheets and tiles of PVC, thermoplastic, cork and rubber. The flooring materials were tested by being laid in the access ways to ticket offices on the London Underground, over a period of 4-6 weeks. The site tests were intended to assess abrasion resistance measured by loss of thickness. The differences in rates of wear on all sites between the 11 materials tested were small, with the exception of the 6.4 mm thick cork tiles in which the loss of thickness was very much greater than for the other materials.

Another useful publication is Buildings Research Establishment Digest 33 (revised in 1971): Sheet and tiles flooring made from thermoplastic binders. This makes a number of important points which include the following:

Adhesives must be carefully selected as some can cause the PVC flooring to shrink; this is caused by the movement of the plasticizer from the flooring into the adhesive, and the amount of shrinkage can be high, up to 0.7%. The plasticizer softens the adhesive and the flooring moves under traffic. Another problem is the movement of compounds from some types of adhesive, e.g. tar based, upwards into the flooring resulting in unsightly stains.

PVC tiles for industrial floors

A patented PVC tile which has been used successfully for warehouses and other industrial use is the Altro Hyload silica-quartz-reinforced PVC tile and 'plank'. The tiles are square (300 mm × 300 mm) and the planks are rectangular (300 mm × 600 mm).

When laid on ground-supported concrete slabs, an effective damp-proof course (dpc) is essential. It is recommended that the adhesive used should be supplied by the manufacturer of the tiles. The moisture content of the base on which the tiles are to be laid should give a hygrometer reading of not more than 75%; for details of this dampness test see Appendix A of BS 8203: Code of Practice of installation of sheet and tile flooring.

The manufacturers attach considerable importance to their requirement that the temperature of the sub-floor should not be lower than 13°C and the temperature of the PVC tiles should not be lower than 18°C when laying is to take place. It is recommended that the detailed instructions of the manufacturers are carefully followed.

Figure 4.5(b) shows a floor finished with these Hyload tiles.

Concrete base slabs with timber flooring

Introduction

Concrete base slabs finished with timber are popular in domestic buildings and in some sections of commercial buildings, laboratories and in certain industries. This section does not deal with suspended timber floors, but only those in which the timber is either bedded directly onto the concrete/screed or on bearers which themselves are bedded on/in the concrete/screed or fixed to the base slab.

General considerations

The moisture content of the timber-supporting battens and the floor boarding and blocks must be strictly controlled at the time of fixing and all necessary precautions taken to ensure that the moisture content in use does not exceed recommended values. The following maximum moisture contents are recommended by the Building Research Establishment and in the relevant Code of Practice (BS 8201):

- *floor battens*: 22% at the time of fixing, which should reduce quite quickly in use to a maximum of 20%;
- *floor boarding and blocks*: at the time of fixing, not more than 2% higher than the maximum recommended in use, which will be determined by heating conditions as set out below:

– without any artificial heating	15–19%
– with intermittent heating	10–14%
– with continuous heating	9–11%
– with underfloor heating	6–8%

The moisture content of timber can be checked on site by moisture meters of the conductivity type. The reading can be affected by the presence in the timber of preservatives and in such cases the manufacturers of the meters should be consulted.

Timber properly fixed at the correct moisture content can be adversely affected by external moisture from three sources:

(a) water spilt on the floor surface;
(b) water rising from the ground (ground-floor slabs only);
(c) residual moisture in the screed or concrete with which the timber is in close contact.

During construction, precautions should be taken to protect the timber flooring from spillage of liquids, and users should be warned of the probable effects (swelling and lifting) if the timber flooring becomes wet.

In the case of water rising from the ground (ground-supported slabs only), this should be taken care of by the provision of an effective damp-proof membrane as required by the Building Regulations 1991 and Approved Document C. Some detailed information on dpms in ground-floor slabs has been given in Chapter 2.

Figure 4.6 shows a typical detail suitable for timber flooring on ground-floor concrete slabs.

The following types of wood or wood-based flooring are in general use:

- wood blocks (to comply with BS 1187);
- timber board or strip flooring;
- panel flooring:
    flooring-grade plywood (to comply with BS 6566),
    flooring-grade chipboard (to comply with BS 5669, with special reference to Part 2),
    cement-bonded particle board (to comply with BS 5669, with special reference to Parts 4 and 5),
    mosaic parquet panels (to comply with BS 4050).

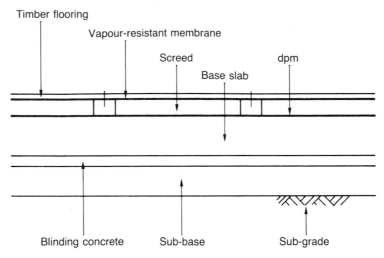

Fig. 4.6 A sketch showing a concrete base slab with timber flooring.

It is of particular importance to ensure that the plywood, chipboard and particle board can be obtained in a grade suitable for its intended use.

As far as wood blocks are concerned, it is important to note that they are now usually laid on a cold bitumen/rubber adhesive rather than hot-applied bitumen or pitch. According to the Building Research Establishment, the adhesive should not be relied upon to provide the essential damp-proof membrane; this means that a dpm must be provided separately, either below the screed on top of the base slab or if a screed is not provided, below the concrete slab on a suitable thickness of over-site concrete.

If the boarding, strip flooring or panels are to form the wearing surface, then the following factors should be taken into account, depending on the duty the floor has to perform: resistance to wear, resistance to dimensional change, appearance, slip resistance, fire resistance, and sound and thermal insulation.

A list of suitable timbers for a wide variety of uses is given in Table 1 of BS 8201: Code of Practice for flooring of timber.

Access floors (platform floors)

For present day conditions the internal layout of buildings in commercial occupation needs to be as flexible as possible so that changes in user requirements can be accommodated with minimum cost and dislocation, with special reference to the need to accommodate electric cables ducting and pipe-work. One answer to this need has been the development of access flooring systems.

An access floor is essentially a prefabricated floor surface (panels), supported on pillars with a usable and easily-accessible void between the underside of the panels and the surface of the structural floor slab. Figure 4.7 is a diagram of an access floor system.

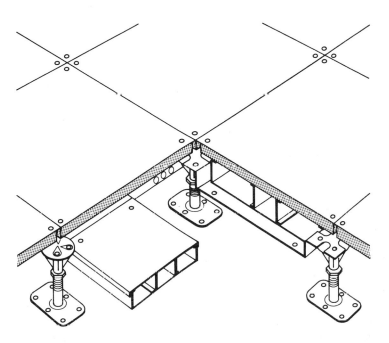

Fig. 4.7 A diagram of a raised access floor. Courtesy: Thorsman & Co. (UK) Ltd.

There are now many access floor systems on the market. Some have been granted Agrément Certificates; some suppliers are Quality Assured Firms under BS 5750: Part 2. They all claim that their particular system meets the Building Bye-Laws for spread of flame (Class '0'), meets the requirements of PSA Standard MOB.08.801 (which is now obsolete and has been replaced by MOB.PF2.PS), and is a comprehensive Performance Specification for Platform Floors (Raised Access Floors) of some 92 pages.

Important matters to be considered in evaluating an access flooring system include the following:

(a) Compliance with recognized Standards, with particular reference to the resistance to spread of flame (Class '0'), structural stability and resistance to vibration and lateral movement (rigidity), and durability under operating conditions.

(b) The panels and supports must be able to carry the design live and dead load with minimum deflexion; the PSA specifications require this to be a maximum of 1/250 of the shortest span or 2.5 mm, whichever is less.

(c) A low dead (self) weight.

(d) Maximum accessibility and flexibility for the accommodation of services.

(e) The design must allow for very fine adjustment in the level of the panels so that the finished floor is level to within fine tolerances which should be specified.

(f) The floor outlet units must be functional, easily installed and readily repositioned.

(g) The floor panels must be securely fixed but be readily removable.

The following comments are also relevant:

(a) When a pillar/column support system is used, it is important to ensure that the substrate will support the point loads from these supports. When the substrate is the concrete base slab there should be no problem. However, if the supports are carried on a cement/sand mortar screed, then the screed should be specified to withstand the BRE impact screed test. The depth of indentation should be related to the load carried and the specified factor of safety.

(b) The substrate must be suitable for very secure fixing of the column bases (whether by holding down bolts or adhesive). The PSA specification contains test requirements for the fixing of the panels to the supports and the fixing of the supports to the substrate.

(c) As the access floor is designed to allow fine adjustment in the level of the panels, there is no need to specify very close tolerances on the finished levels of the structural slab nor of the screed.

(d) The panels are generally high-density particle board to BS 5669, encapsulated in a zinc-coated steel sheet. Other types of panel are available.

(e) In several systems the vertical supports are circular pillars of cast aluminium or steel, with a screw thread on the outside which allows very fine adjustment of the level of the floor panel. One system uses aerated concrete blocks with inert PVC shims for fine levelling.

(f) An Italian system uses 600 mm × 600 mm × 11 mm ceramic tiles for the finished surface of the panels; the tiles being bonded to a support panel 34–40 mm thick, depending on the material used. The support structure consists of circular galvanized steel columns.

(g) Specifiers should study carefully the performance specification prepared by the PSA, reference MOB.PF2 PS. 1990.

Sports hall floors

The relevant British Standard is BS 7044: Parts 1 and 2: Artificial sports surfaces.

Part 1: Classification and general introduction
Part 2: Methods of test (Sections 2.1 to 2.5)

The Standard covers both indoor and outdoor sports surfaces, and points out that, ideally, each sport activity should have the surface individually specified for that particular sport. For economic reasons it is usual to find several different sports played on the same surface. However, it must be kept in mind that even if a particular surface appears to be satisfactory for a given sport as regards playing characteristics, it may not be satisfactory in terms of safety. When in doubt, expert advice from the Sports Council should be sought and followed.

It is also advisable that the playing surface should be selected as early as possible in the design process, so that the recommendations of the material suppliers can be given full consideration as far as the floor as a whole is concerned. The brief comments given in this chapter deal with floors of multi-purpose sports halls where the playing surface is laid on a concrete base slab, and only two types of playing surface are described: composition blocks and various grades of sheet vinyl.

Composition block floors

The use of composition block flooring is covered in a general way by BS 5385: Part 5. These blocks are all proprietary materials and therefore the recommendations in the Standard are necessarily restricted. The Standard states: 'A typical block is a mixture of cement, wood granules, mineral fillers, pigments and water.' The recommendations in the Code are supplemented by detailed directions given by the suppliers of the blocks, who usually appoint specialist contractors to lay the blocks.

One of the best known firms making composition blocks for sports hall floors is the Granwood Flooring Company, Derby. The Granwood blocks are considered a 'breathing' floor, so that a damp-proof membrane is not required immediately below the blocks as the blocks are normally able to cope with the evaporation of moisture from the screed/concrete floor slab. However, when the floor is in contact with the ground (a ground-supported floor slab), a dpm is normally considered necessary to comply with the Building Regulations 1991 and Approved Document C (1992 edition).

The nominal dimensions of the Granwood blocks are 174 mm × 57 mm × 10 mm thick. There are two basic types of Granwood flooring, the 'Traditional Granwood' which is laid directly onto cement/sand screed and the 'Gran-sprung'. In the Gran-sprung floor, the composite blocks are bonded to two layers of exterior-grade plywood sheets, each 12 mm thick, which are supported on a large number of 8 mm thick resilient pads. The pads are laid on 1000 gauge polythene sheet on the concrete base slab. Both types of flooring comply with classification HD/1 Heavy Duty Indoor Use, in BS 7044: Part 1.

Figure 4.8 is a view of a multi-purpose sports hall floor finished with Granwood blocks.

Fig. 4.8 View of a composition block floor in a multi-purpose sports hall. Courtesy: Granwood Flooring Ltd.

PVC (vinyl) surfaces

The basic requirements for PVC flooring have already been summarized in this chapter under the section on flexible sheet and tile flooring.

Suitable foam-backed sheet vinyl is made by a number of firms. The following comments are taken from the technical leaflets issued by Gerland Limited, London for their 'Taraflex' range of PVC foam-backed sheeting (see Fig. 4.9).

Fig. 4.9 View of a multi-purpose sports hall finished with Gerland Taraflex vinyl flooring. Courtesy: Warner Group; photo by Warren Denyr Design.

The quality of the finished floor will only be as good as the sub-floor over which it has been laid. All such flooring materials require a smooth, hard, clean and even surface for satisfactory adhesive bond and resistance to wear if long-term durability and satisfactory performance are to be achieved. The moisture content of the concrete/screed at the time of laying the flooring must not exceed the limit required by the supplier of the sheeting. With ordinary concrete and cement/sand screeds, a drying-out time of about 1 day for each millimetre of thickness can be used as a guide only. The maximum permitted moisture content of the concrete or the cement/sand screed and the method of measuring it should be agreed with the suppliers of the sheeting and written into the contract specification. Unless this is done, serious problems/disputes can arise on site over the delay caused by waiting for the concrete screed to 'dry out'.

If the sheeting is laid on a screed then it is essential that it should pass the BRE impact test for heavy duty floors (Category A use, as in Tables 1 and 3 in BS 8203). The interpretation of the test results should follow the recommendations in BRE publication IP11/84: BRE Screed tester: classification of screeds, sampling and acceptance limits.

The Taraflex range of sheeting varies in thickness according to the use to which the floor will be put; for multi-purpose sports halls, the thickness recommended is 6.2 mm. The sheets are about 30 m long and 1.5 m wide. All the joints are hot welded using Taraflex welding rods.

Bibliography

Publications and papers detailed in the text are not included here.

BS 8000: Part 9 Workmanship on construction sites, Code of Practice for cement-sand floor screeds and concrete floor toppings, British Standards Institution, Milton Keynes

Building Research Establishment (1979) *Estimates of Thermal and Moisture Movements and Stresses: Parts 1, 2 and 3*: Digests 227, 228 and 229, July to Sept. 1979.

Harrison, R. and Malkin, F. (1983) On-site testing of shoe and floor combinations, *Ergonomics*, **26**(1), pp. 101-108

ICI (1989) Technical publication on Ucrete industrial floors, p. 12 Thoro System Products Ltd, Redditch

James, D.I. (1984) *Slip Resistance Tests for Flooring, Two Methods Compared*, Rubber & Plastics Research Association (RAPRA), Members Report No. 94, p. 16 Shawbury, Shrewsbury

National Federation of Terrazzo, Marble and Mosaic Specialists Handbook, 1992, p. 40.

Warlow, W.J. and Pye, P.W. (1974) *The Rippling of Thin Flooring over Discontinuities in Screeds*, Building Research Establishment Current Paper 94/74, Oct. 1974, p. 7

Chapter 5
External concrete paving

Introduction

External concrete paving can be *in-situ* or precast, reinforced or unre-inforced. All types must be resistant to frost and some to de-icing salts. Air-entrained concrete should be used where appropriate and recommended in this chapter. Readers should refer to Chapter 1 for information on air-entraining admixtures.

This chapter covers:

(a) garden and landscape paving;
(b) parking areas for private and light commercial vehicles, includ-ing pedestrian precincts;
(c) hard standings/parking areas, short service roads to industrial premises including warehouses, and petrol service stations.

Paving in special areas such as ports and container depots, which have to carry exceptionally heavy axle loads, are not dealt with in this book, but for interested readers, a number of references are given in the Bibliography at the end of this chapter. Roads, airport runways and taxiways are also not included.

External concrete paving, especially when used as parking areas, can become rather badly stained with petroleum oils. These do not attack concrete, but detract from its aesthetic appearance. The application of a high-quality surface sealant, such as Lithurin (based on magnesium silico fluoride) will help to reduce these disfiguring stains.

Landscape paving

As landscape paving is often laid by do-it-yourself enthusiasts, the fol-lowing recommendations are intended as a guide to good practice; for practical reasons, reinforcement and air entrainment is not suggested.

The final level of paving close to buildings must be at least 150 mm below the level of any damp-proof course (dpc) in the walls of the building.

Vegetable matter and top soil should be stripped down to the previously-decided level relevant to the level of the surface of the paving.

On clay, peat or filled ground, it is better to lay about 75 mm compacted thickness of gravel or similar granular material. If brick hardcore is used it must be broken up into pieces not larger than 50 mm, well compacted and 'blinded' with a thin layer of sand.

On the natural ground or on the compacted sub-base, one should lay 500 gauge polythene sheeting to act as a slip layer and to prevent loss of moisture from the concrete.

The concrete should be laid within side-forms, which can be removed about 2–3 days after laying the concrete. The surface of the side-forms in contact with the concrete should be coated with a release agent to prevent the concrete bonding to the forms; a light motor oil can be used for this. The concrete should be laid to a minimum thickness of 75 mm; 100 mm is better as it allows for irregularities in the ground or sub-base. The maximum bay size is 2 m × 3 m.

For site-mixed concrete which is practical for small areas, the following mix proportions, by volume, are recommended:

1 bag of cement (50 kg) – this is about 1¼ cubic feet (0.036m^3)
3 cubic feet (0.086m^3) of sand
4½ cubic feet (0.128m^3) of coarse (20 mm) aggregate

This mix is about 1:6 by volume (cement: aggregate) If 'all-in' aggregate is used, then 1 bag of cement to 7 cub. ft. of aggregate would be an appropriate mix. Sufficient water should be added to make a workable mix which can be easily spread and compacted and finished.

As soon as possible after finishing, the concrete must be covered with polythene sheeting, well lapped and held down around the edges with bricks, blocks or scaffold boards, and kept in place for at least 4 days; this is to cure the concrete, i.e. to reduce early-age loss of moisture.

The concrete should be laid in bays; square bays if possible but if not, the length should not exceed 1½ times the length.

The surface, particularly on a slope should be non-slip (see Fig. 5.1).

Many users of concrete want a pigmented concrete because they do not like the normal grey colour of concrete. Pigments have to be added to the concrete mix on site and this should be done before water is added. The amount of pigment should be decided by trial mixes, and once decided, the amount of pigment used must be carefully controlled. Even with great care, the results can be, and often are, disappointing due to variations in the tone (intensity of colour). It is impossible to obtain the same standard of uniformity of colour as is

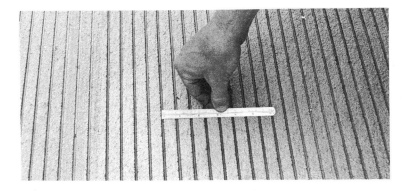

Fig. 5.1 View of in situ concrete paving grooved to provide a non slip surface.

given by a pigmented coating. These unavoidable variations in pigmented in-situ concrete arise from inevitable variations in the mix proportions, standard of mixing, amount of water in the mix, and variations in compaction and finishing.

In factory-made concrete products, the amount of variation is much reduced but not completely eliminated. Pigments used should be specified to comply with BS 1014: Pigments for Portland cement and Portland cement products.

Precast concrete paving blocks can be laid to attractive patterns, but the work is best carried out by an experienced firm; the blocks should comply with BS 6717: Part 1, and they should be laid in accordance with Part 3. Figure 5.2 shows a paved area in a landscaped garden.

Garage drives

The laying of an in-situ garage drive is a bigger undertaking than laying garden paths, and it is likely that the volume of concrete required will justify the use of ready-mixed concrete, and the work being carried out by a local contractor.

The order to the supplier should state the purpose for which the concrete is to be used, and be based on BS 5328: 1990: Designated mix PAV 1, to Table 13 of Part 1 and Table 7 of Part 2. This would take care of the characteristic strength (35 N/mm^2), the minimum cement content (300 kg/m^3), the maximum free water/cement ratio (0.60), the slump (75 mm), the 20 mm maximum size aggregate, and the air entrained (mean 7% entrained air). (See Chapter 1 for comments on air-entraining admixtures.)

A thickness of 150 mm is recommended, laid on a slip membrane on a compacted sub-base at least 100 mm thick. The slab should be reinforced unless it is laid in bays not exceeding 4 m × 4 m. The use of

Fig. 5.2 Area surfaced with concrete block paving in a landscaped garden. Courtesy: Redland Precast Ltd.

fabric reinforcement, such as C385 mesh (to BS 4483), located 50 mm below the top surface of the slab, would allow the bays to be 10 m long and 4.5 m wide.

It is assumed that the transverse joints at 10 m centres will be stop-end joints with the reinforcement stopped back 75 mm each side of the joint. An attempt to saw cut the joints or wet form them in the correct position could lead to practical difficulties on site with this class of work.

The concrete should be laid between side-forms, well compacted and properly cured for 4 days by covering with polythene sheets or by the application of a sprayed-on resin-based curing compound.

Pedestrian precincts and parking areas for private cars and light commercial vehicles

In-situ reinforced concrete

These areas are likely to be reasonably large and the author recommends that the in-situ reinforced concrete paving should be designed, specified and constructed on the same principles as those for paving for heavy commercial vehicles, but allowing for less heavy loading. It is assumed that the work will be carried out by experienced contractors under reasonable site control.

Each case should be considered separately, but the following is intended as an illustration:

– Sub-grade: sandy clay, CBR 8.
– Sub-base: granular material: 200 mm thick.
– A slip membrane of 1000 gauge polythene sheeting should be laid on the compacted sub-base, lapped 150 mm.
– Concrete slab: 180 mm thick.
– Bay length between free movement joints: 50 m.
– Reinforcement: fabric-B385 (BS 4483) providing 385 mm² in the longitudinal wires and 193 mm² in the cross wires. This would allow the bay width to be up to about 10 m.
– Concrete for the slab can be a designated mix from BS 5328, namely, PAV 2 from Table 13 of Part 1 and Table 6 of Part 2. This takes care of all essential factors relating to the mix, including air entrainment. Figure 5.3 shows an enlarged detail of a cut section of air-entrained concrete.

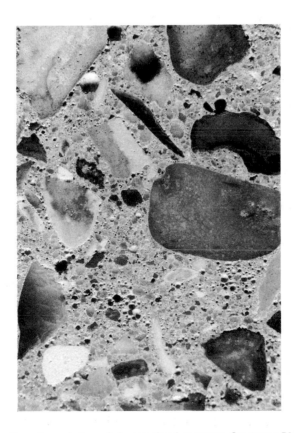

Fig. 5.3 A greatly enlarged view of air-entrained concrete. Courtesy: RMC Ltd.

Stress-relief joints should be sawn or wet formed at 10 m centres and should extend down to a depth of 1/4 to 1/3 of the depth of the slab.

The steel fabric should have a cover of 50 mm ± 5 mm for slabs less than 200 mm thick, and 60 mm ± 10 mm for slabs 200 mm or more in thickness. The reinforcement should be stopped off for a distance of 75 mm each side of the joint.

In the few cases where it is certain that only private cars and pedestrians will use the area, bottom reinforcement across the induced joints can be omitted. This means that reliance is placed on aggregate interlock to provide effective load transfer across the induced joint (which is in fact a crack that extends the full depth of the slab, induced by curtailing the top reinforcement and sawing or wet-forming the joint as described above).

However, the author feels that it is generally prudent to provide bottom reinforcement and not to rely on aggregate interlock for load transfer. The bottom reinforcement should consist of the same mesh as in the top of the slab, placed so that the main bars run longitudinally (as in the top mesh). This reinforcement should be securely fixed at a distance of about 1/3 or 1/4 of the slab depth above the slip membrane, and should extend 500 mm each side of the induced joint.

While there are advantages with sawn joints, there are also difficulties in timing the sawing so as to ensure that no damage occurs to the joint edges from too early sawing and premature cracking of the slab due to delayed sawing. Longitudinal joints should be tied with 12 mm diameter steel tie bars, 1.0 m long, at 600 mm centres. As suggested above, the fabric reinforcement is suitable for bays up to 10 m wide, which reduces the number of longitudinal joints; however, in this case the tie bars should be debonded for half their length, with care being taken to ensure correct location at half-slab depth and that all debonded lengths are on the same side of the joint.

If narrower bay widths (say 4.5 m) are selected, then the tie bars need not be debonded, except at every third or fourth longitudinal joint.

Debonded dowel bars can be used instead of debonded tie bars, and these should be 25 mm in diameter, and 600 mm long, located at 300 mm centres. The author suggests that dowel bars should only be used in slabs 200 mm or more in thickness. In thinner slabs, even a small displacement of the dowel bars can result in crack formation. Even with the thicker slabs, significant shortcomings in assembly are likely to result in cracking of the slab which would probably require costly rectification. Due to problems arising from defects in dowel bar assembly, a proprietary double-dowel shear load transfer assembly has been produced. This system was developed in Switzerland and is marketed in the UK by George Clark (Sheffield) Ltd, who offer a complete design service for their product.

What may be termed projections or protrusions into the bays, e.g. gulleys and manhole covers, can cause diagonal cracks, radiating from the corners of the gulley frame, manhole cover, etc. A good arrangement is to box out the gulley or manhole. The joint between the boxed-out area and the adjoining bays should be a plain butt joint with the concrete at the interface debonded, and reinforcement stopped off 50 mm from the joint. Dowel bars or tie bars, debonded on one-half of their length, may be required at manholes located in traffic routes.

It is sometimes necessary to provide irregularly-shaped areas to fit in with entrances and exits but, with care, triangular-shaped bays can be avoided.

It is usual to provide a reasonably non-skid surface to the slabs. A brush finish is satisfactory if it is carried out in accordance with the recommendations in the Department of Transport's Specification for Highway Works (7th edn), Series 1000, Clause 1026 and Table 10/6.

On completion of compaction and finishing, the concrete should be cured for 7 days by the application of a resin-based aluminized curing compound.

Cracking in concrete slabs

It is the author's experience that the majority of complaints about defects in concrete pavements arise from cracking. Where air-entrained concrete has not been used, complaints usually arise from spalling of the surface due to the action of freeze-thaw conditions. Water increases by about 20% in volume when it turns into ice, and the resulting expansion produces stresses which disrupt the surface layers of the concrete.

The cracks in concrete may be shallow, extending down to the reinforcement, as in mild cases of plastic shrinkage cracking. Or they may extend down the full depth of the slab and are then considered as structural cracks. The width of the crack should be determined at an unspalled length on the surface of the slab. Cracks up to 0.5 mm wide are considered as narrow; cracks between 0.5 and 1.5 mm wide are classified as medium; and those above 1.5 mm as wide. According to a Department of Transport publication, 'narrow transverse cracks are a normal feature of all reinforced road slabs', and do not require remedial work. Longitudinal cracks should not occur and the same applies to diagonal cracks, i.e. cracks that are neither transverse nor longitudinal.

All early-age cracks arise from restraint on dimensional change in the concrete after it is laid and before any external loads are applied to it. Concrete contracts as it cools down (thermal contraction) and matures. It also contracts as it loses moisture by drying out after the end of the curing period (drying shrinkage). Stress is induced in a newly-placed concrete slab by restraint from adjacent slabs to which it is tied by reinforcement or bond, and by friction at the interface of the slab and the sub-base. The likely cause of the cracking should be investigated as soon as possible after the cracks are reported, which in fact may be some time after they actually occurred.

A type of cracking that can cause considerable consternation is plastic shrinkage cracking, which occurs within the first 12 h after casting and finishing the slab. It is caused by too rapid evaporation of moisture from the surface of the concrete while the concrete is still plastic. When the rate of evaporation of moisture from the surface of the plastic concrete exceeds the rate at which the 'bleed' water rises to the surface, plastic shrinkage cracking is likely to result. There are many factors involved, but the following are known to play an important part:

- relative humidity, wind velocity, temperature of the air, degree of exposure of the concrete to the wind and sun;
- temperature of the concrete;
- cement content and total quantity of water in the mix;
- grading, absorption and type of aggregate used;
- type of admixtures used (if any);
- density of the compacted concrete;
- thickness of the slab.

Plastic shrinkage cracking appears as fine cracks which are often fairly straight and more or less parallel to each other; the length generally varies from 50 mm to 750 mm, the distance apart varies from 50 mm to 90 mm, and these cracks are often transverse in direction. The cracks are usually shallow and seldom penetrate below the top layer of reinforcement. In severe cases they can extend right through the slab.

Figure 5.4 shows a severe case of plastic shrinkage cracking, which required the replacement of the bay. A simple method of repair to 'normal' plastic shrinkage cracks is to brush in a cement-based polymer-modified grout as soon as possible after discovery and protect the surface from wind and sun. Methods of crack investigation and repair are discussed in Chapters 6 and 7.

Precast concrete slabs/flags

For ornamental paving there is a wide variety of shapes, sizes, colours and textures on the market, including what is known as 'reconstructed/reconstituted' stone. The thickness varies from about 40 mm to 65 mm, but for sidewalks the minimum thickness is 50 mm and the flags should comply with BS 7263: Part 1: Specification, and Part 2: Code of Practice for laying. This Standard replaces BS 368. The

Fig. 5.4 Plastic cracking in the surface of a concrete slab. Courtesy: A/S Scancem, Norway.

Standard lays down requirements for transverse strength and water absorption, as well as dimensions and tolerances on size and thickness. Sampling must be carried out before the flags are laid. The Standard refers to three methods of manufacture but for the purpose of the Standard, only two methods are suitable, namely the semi-dry and the wet pressed.

Recommendations for laying the flags are given in Part 2 of the Standard to which reference should be made for details. Essentially, the flags are laid on a properly prepared sub-base, and the bedding can be either a mortar or sand/crushed rock fines, graded to C or M, Table 5 of BS 882. The mortar mix can be either cement and sand (1:3) or lime and sand (1:3); the sand grading to be M or F, Table 5 of BS 882. The mortar bed should be a compacted thickness of 25 mm. When the flags are laid on sand or crushed rock fines, the bed should be 30 mm thick. The flags must be fully and uniformly bedded and well punned down. The use of a lime mortar is recommended for coloured flags as this is easier to clean off the flags than a cement-based mortar.

The flags should be carefully stored on edge, on a clean polythene sheet which is laid on timber runners. This is particularly important with pigmented flags.

Complete uniformity of colour will not be obtained with pigmented flags and there is always the risk of 'lime bloom' (calcium carbonate leached from the cement) appearing on the surface of the slabs. This discolouration is particularly noticeable on the darker colours, and dark grey and black should always be avoided. Pigments should be specified to comply with BS 1014: Pigments for Portland cement and Portland cement products.

Joints in the paving can be 'narrow' (i.e. about 3 mm wide) or 'wide' (i.e. about 5–10 mm wide). The narrow joints are filled by brushing in a cement/sand semi-dry mortar. The wide joints are filled by hand with a cement/sand mortar. In both cases the flags should be cleaned as the joint filling proceeds. The mortar mix should be about 1:3.5 by volume. For the wide joints, the mortar can be pigmented so that the joints can be made a feature of the whole layout.

If the flags are laid on filled ground, e.g. to form a walkway around a swimming pool, then it is inevitable that the filling will settle with time and the paving will become uneven, with some flags cracking. This can be easily rectified by taking up the flags, making up the base with sand, and re-laying them to the correct levels. Precast flags thus have a great advantage over in-situ paving. It must be remembered that the colour of new flags will not match closely the colour of the old existing ones.

Heavy-duty concrete rafts

The concrete raft type of paving is suitable for carrying heavy static and moving loads. The best known 'raft' in this specialized field is the Stelcon Concrete Raft, manufactured by Redland Precast Barrow-upon-Soar (see Figs 5.5 and 5.6); the information given below was obtained from Redland.

The Stelcon Raft is a precast unit made of OPC reinforced concrete in a steel angle frame to protect the arrises. The standard sizes are: 1995 × 1995 × 140 mm and 1995 × 1295 × 140 mm. Weights are 1340 kg and 890 kg respectively. The wearing surface is textured granite, but a trowelled surface can also be provided. A steel-clad surface can also be obtained when resistance to severe abrasion is required. The units are laid on a bed of compacted graded sand, which in turn is laid on a sub-base, on a prepared sub-grade. Detailed requirements for the sub-grade, sub-base and sand bed will depend on the loading conditions, and advice on this should be obtained from the suppliers of the units.

Suitable uses for these units include access roads to factories, hard-standing areas, docks, transit sheds and warehouses. The advantages are:

- rapidity of construction and the ability to take traffic immediately after laying;
- high load-carrying capacity;
- in areas where settlement of the sub-soil/compacted fill is anti-

Fig. 5.5 A view of Stelcon Rafts being laid. Courtesy: Redland Precast Ltd.

Fig. 5.6 A view of a completed area of Stelcon Rafts for heavy duty use. Courtesy: Redland Precast Ltd.

cipated, the units can be taken up and re-laid to corrected levels; they also provide easy replacement should damage occur.

Concrete block paving

Concrete block paving has become very popular in the UK during the past 15 years, but has been used extensively on the Continent for many years.

There is now a British Standard for paving blocks: BS 6717: Part 1, which is a specification for the blocks themselves, and Part 3, which is a Code of Practice for laying the blocks. The following is an outline of the basic recommendations for laying block paving in areas carrying light to medium traffic. Details can be obtained from the relevant trade organization, namely, Interpave, which is part of the British Precast Concrete Federation. Reference should also be made to the Code of Practice, and to BS 7533: Guide for the structural design of pavements constructed with clay and concrete block pavers. The blocks are not air entrained, but have very high strength (a crushing strength of 49 N/mm^2) and a minimum binder content of 380 kg/m^3 which makes the blocks very resistant to de-icing salts, and to abrasion from traffic.

Pigmented blocks can be obtained in a limited range of colours.

The sequence of operations is as follows. The ground has to be excavated to the required level. A properly constructed sub-base must be specified and the details are dependent on the sub-grade. For example, with a sandy clay, a water table more than 600 mm below formation level and a CBR of 6%, a sub-base of selected and compacted granular material of about 180 mm thick should be adequate. More detailed information on design is given in the publications listed in the Bibliography at the end of this chapter.

The blocks are laid on a 50 mm thickness of compacted sharp, clean sand; but see Part 3 of BS 6717, clause 3.1 which recommends that construction generally should be in accordance with the Department of Transport's Specification for Highway Works 1986, except where varied by the Project Specification. It should be noted that the Department of Transport's Specification is now in its 7th edition, published in 1992.

BS 6717 lays down strict requirements for the grading of the sand on which the blocks are bedded and also for the sand for filling the joints. Recommendations are given gradients, namely a cross fall of 1 in 40 and a longitudinal fall of 1 in 80. Tolerances are given on surface levels. Edge restraint is essential to prevent the blocks moving from their as-laid positions; precast concrete kerbs complying with BS 7263 are very suitable for this purpose.

The blocks are laid by hand and suitable areas are vibrated by a plate vibrator as the work proceeds, but vibration should not be carried out near an unrestrained edge. Sand is then spread over the surface and brushed into the joints, and the plate vibrator is again passed several times over the area. The paving is then ready for use.

Although the above description may suggest that an area paved with concrete blocks is self-draining, this is in fact not the case and drainage should be provided, which requires that the whole area must be laid to gradients and discharge to a satisfactory surface water drainage system.

Figure 5.7 shows an area of block paving carrying heavy loads.

Fig. 5.7 A view of concrete block paving. Courtesy: Redland Precast Ltd.

Drainage for paved areas

The comments which follow apply to all types of paving.

Recommendations for the design of paved area drainage, based on the principles of hydraulics and local meteorological knowledge, are contained in BS 6367: Code of Practice for drainage of roofs and paved areas. The details of surface water drainage design and construction is outside the scope of this book.

A drainage system for a paved area normally consists of a channel or channels which collect the run-off from the paving, and these channels discharge to road gulleys located at predetermined positions; the gulleys are connected to a surface water drainage system which is usually underground. This 'drainage system' discharges either to a surface water sewer, a combined sewer, a water course of to one or more soakaways. The design and construction of soakaways need careful consideration and reference should be made to BRE Digest 365, September 1991.

The paving must be laid to a fall/gradient, and a generally accepted gradient is 1 in 60 for the paving and a minimum of about 1 in 150 for the channels. If special channel blocks are used, the gradient can be reduced to about 1 in 200. The spacing of the road gulleys for paved areas depends on the run-off, the capacity of the collecting channels that will discharge to the gulleys, and the flow capacity of the gulley inlets.

The levels of the paving should take into account:

– levels of the existing ground;
– levels of access-ways to the paved area;
– levels of existing or proposed surface water drainage pipes/ channels/outfalls, or suitable locations for soakaways;
– levels of the damp-proof course in any buildings adjacent to the paving.

Heavily-loaded paving

In-situ reinforced concrete

In-situ reinforced concrete is very suitable for paving that has to carry heavy commercial vehicles. Reference should also be made to 'Fast Track' concrete paving in Chapter 7.

The principles of design and construction are similar to those used for roads. The main difference in use between the external paving dealt with in this book and roads is that roads carry vehicles travelling at a much higher speed.

For many years the basic document for the design of concrete roads in the UK was Road Note 29 issued by the Department of the Environment Road Research Laboratory in 1960, revised in 1965, with a 3rd edition in 1970. The recommendations in Road Note 29 were mainly based on what were termed 'experimental roads' and the information obtained from monitoring these roads was valid up to 1970. According to the publication from the Transport and Road Research Laboratory in their Research Report 87, dated 1987, none of the roads covered by Road Note 29 had carried more than 10 million standard axles (10 msa).

The Research Report 87: Thickness design of concrete roads, is based on observations of the original 'experimental roads' after 1970 together with other experimental sites so that the research is based on the effect of a much greater volume of traffic on the roads in question.

Research Report 87 was given formal design recognition by the publication of the Department of Transport's Departmental Standard HD 14/87 and Advice Note HA 35/87. Complete bibliographical reference to these publications is given in the Bibliography at the end of this chapter.

It should be noted that the latest issue of the Department of Transport's Specification for Highway Works is the 7th edition, published in 1992. In this document, concrete pavements are covered by Series 1000.

It appears that original design conceptions for concrete pavements are attributed to H.M. Westergaard in the early 1920s and were published in the *Proceedings of the 5th Annual Meeting of the Highway Research Board*, Vol. 5, Part 1, pp. 90–112, in 1925. As far as the USA and Canada are concerned, reference should be made to the publications of the following authorities:

American Association of State Highway and Transportation Officials (AASHTO)
The American Concrete Institute
The Portland Cement Association (USA)
The Ministry of Transportation and Communications, Ontario
The Portland Cement Association (Canada)

Essentially, the design of the pavement will depend on:

- the estimated traffic load, defined as the number of equivalent standard axles (80 kN);
- the design life, expressed as cumulative standard axles; usually referred to as 'million standard axles (msa);
- the type and California Bearing Ratio (CBR) of the sub-grade;
- the type of sub-base selected;
- the thickness of the concrete slab;
- the area of reinforcement per metre width of slab provided in the longitudinal direction and, int the case of wide bays, in the transverse direction;
- the length of the slab between 'frec' movement joints.

In-situ concrete paving reinforced with random steel and polypropylene fibres

The use of randomly-dispersed fibres (steel and polypropylene) is more common for industrial floor slabs than for external paving, but the author believes they can also be used successfully for the latter purpose. The fibres can be used either on their own as is recommended by Euro Steel, Brussels for their special steel fibres, or polypropylene fibres such as 'Crackstop' supplied by Castle Building Products, Hawarden or 'Fibrin' supplied by Fibrin (Humberside) Ltd, or 'Fibremesh' supplied by Fibremesh Europe Ltd, Chesterfield. These are all proprietary materials and, as is natural, each supplier makes particular claims about their own product.

The absence of BSI, ACI and ASTM specifications and test methods for the fibres and for the fibre-reinforced concrete (FRC) restricts the acceptance by Consulting Engineers and paving designers to the use of fibre-reinforced concrete for both industrial floors and external paving. Generally, the use of the random fibres is claimed to reduce the incidence of shrinkage cracking and improve other characteristics of the concrete, such as an increase in cohesion, impact resistance, and a reduction in slab thickness. The polypropylene fibres can be used instead of standard steel fabric or in addition to such reinforcement. There is a tendency for a small number of these fibres to project above the finished surface of the slab, but these fibres will disappear under wear.

Steel fibres are usually used without the inclusion of steel fabric (see comments in Chapter 2). If steel fibres are used, and unless they are made of stainless steel or are specially protected against rusting, the surface of the concrete may fairly quickly show brown staining caused by the rusting of a small number of fibres close to the surface of the slab. This rusting has no deleterious effect on the structural capacity of the slab and its effect is only aesthetic. The author considers that as the surface of external concrete will rapidly become stained with oil, tyre marks, mud and leaves, the presence of the brown rusting is unimportant and can be ignored.

Reinstatement of trenches in paved areas

The back-filling of trenches, cut either before or after the construction of external paving, always presents some problems as it is virtually impossible to compact adequately the back-fill, i.e. compact it so that a further reduction in volume does not occur at a later date.

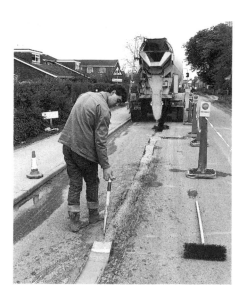

Fig. 5.8 A view of a trench being reinstated with foamed concrete. Courtesy: British Cement Association.

In recent years interest in the long-term reinstatement of trenches in public highways has increased considerably, largely as a result of the Horne Report. Materials that have been found to perform very well as back-fill are dry-lean or wet-lean concrete and foamed concrete; but it is essential that the compressive strength should not exceed 8 N/mm², and the minimum strength varies from 2 to 4 N/mm² according to the surface loading by vehicles.

Figure 5.8 shows a trench being back-filled with foamed concrete.

Bibliography

Publications and papers detailed in the text are not included here.

Dean, R.C., Havens, J.H., Rahal, A.S. and Azevedo, W.V. (1980) Cracking in concrete pavements, American Society of Civil Engineers, *Transport Engineering Journal*, March, 155-169

British Cement Association (1991) Foamed Concrete publications BCA, Crowthorne

British Cement Association (1991) *Plastic Cracking in Concrete*, 2nd edn, BCA, Crowthorne, p. 4

British Cement Association (1992) Concrete pavements for highways, *BCA Bulletin*, July, BCA, Crowthorne, p. 4

BS 6717 Precast concrete paving blocks – Parts 1 and 3, British Standards Institution, Milton Keynes

BS 5489 Road lighting – Part 3: subsidiary roads and pedestrian areas, British Standards Institution, Milton Keynes

Brown, B.V. (1982/1983) Air-entrained concrete, Parts 1 and 2, *Concrete*, Dec. 1982 and Jan. 1983

Cement and Concrete Association (1964) *Bay Layout for Concrete Roads*, Db24, CCA, Crowthorne, p. 16

Department of Transport, New Roads and Street Works Act 1991 and regulations made thereunder, HMSO

Department of Transport, Transport and Road Research Laboratory, and Cement & Concrete Association. A guide to concrete road construction, 3rd edn, HMSO, London, p. 82

Department of Transport (1986/1988) Specification for highway works, Parts 3 and 7, Appendix L, HMSO, London

Department of Transport (1989) *Structural Design of New Road Pavements*, HD 14/87, Dec. 1989, with amendments, HMSO, London

Department of Transport (1992) *Specification for Highway Works*, 7th edn, HMSO, London

Hunt, J.G. (1969) *Curing of Concrete Pavement Slabs in Hot Weather*, Cement & Concrete Association Report 42.435, Nov. 1969, p. 14

Hunt, J.G. (1972) *Temperature Changes and Thermal Cracking in Concrete Pavements at Early Ages*, Cement & Concrete Association Report 42.460, April, 1972

Interpave-Concrete Block Paving Association (1990) Technical literature and information sheets $^c/_o$ BPCR, Leicester

Lilley, A.A. (1991) *Handbook of Segmental Paving*, E. & F. Spon

Lilley, A.A. and Walker, B.J. (1978) *Concrete Block Paving for Heavily Trafficked Roads and Paved Areas*, Cement & Concrete Association Report 46.023, p. 15

Mayhew, H.C. and Harding, H.M. (1987) *Thickness Design of Concrete Roads*, TRRL Research Report 87, Department of Transport, London p.13

Mildenhall, H.S. and Northcott, G.D.S. (1987) *A Manual for the Maintenance and Repair of Concrete Road*, Department of Transport and Cement & Concrete Association Report, HMSO, London, p. 80

Panarese, W.C. (1992) Fiber: Good for the concrete diet? *Civil Engineer (USA)*, May, 44–47

Renier, E.J. (1987) Concrete overlays challenge asphalt, *Civil Engineer (USA)*, April, 54–57

Chapter 6
Notes on contract specifications, site control, investigations and testing

Introduction

The notes which constitute this chapter are essentially brief because a complete book could be written on each of the four subjects covered.

The author's experience is that a badly written, inadequate or technically-unsound specification can make site control difficult and result in a construction dispute involving investigations and testing. A further cause of such a development is the insistence by building owners on the acceptance of the lowest tender. Tenders are often invited on a selected list prepared by the engineer or architect. The principle of such a list is that the person preparing it believes from experience that the firms on it are capable of carrying out the work involved satisfactorily. There are many reasons why one or more of the firms invited to tender submit bids that the experienced professional person knows are so low as to virtually preclude any profit and may result in a substantial loss. The adjudication of the contract to such a firm is a recipe for serious trouble on site. It may well result in the contractor going into liquidation, and the final cost of the project being far higher than if the highest tender has been selected, quite apart from a serious delay in the completion of the project and consequent serious financial loss to the building owner.

Defects appearing in a floor once the building is occupied can prove to be a very serious and potentially costly business, particularly in commercial and industrial premises. Small-scale repairs can usually be carried out over a weekend, but major remedial work requires the temporary closing down of the business with high consequential loss.

Preparation of the specification

The author considers that the specification is one of the basic documents in a contract. His experience suggests that it would be advantageous if architects followed the same procedure as civil and structural engineers by preparing the specification as a separate contract document rather than allowing it to be included as 'descriptions in the bills'.

The basic requirements for a specification are admirably set out in the British Standards Institution publication: PD 6472: Guide to specifying the quality of building mortars. Clause 3 of this document states:

> 'A specification should be free of vague and non-committal phrases . . . because a contractor should not be held responsible for complying with a requirement that the specifier has not clearly expressed. Some characteristics such as colour . . . cannot easily be specified in words. Where they are important, preliminary trials should be demanded as part of the contract and agreed limits of variation determined in relation to reference samples. The specification should be drafted with a clear and realistic concept of the way it can be enforced. As a specification should be a document expressing intent, where possible it should include a statement of the risks the contractor faces in the event of non-compliance.'

As far as floors are concerned, the author's experience is that an inadequate or unclear description of the surface finish results in disputes on site and may lead to subsequent litigation; this could be largely avoided if the specifier makes clear to the contractor what he really wants. In a number of cases in which the author has been involved, he has been forced to conclude that the specifier did not know what he wanted, but was convinced that the standard produced by the contractor was unacceptable.

When specifying proprietary materials, particularly on an important project, the specifier should satisfy himself that the materials in question have been used successfully on other similar projects, or that sufficient relevant development testing has been performed with satisfactory results.

The possession of an Agrément certificate is of course a great help when considering the use of new products and systems. However, such certificates should be perused with great care to see exactly what is covered by the wording. The author considers that the details of testing carried out by the British Board of Agrément prior to the issue of a certificate should be available to users, with the agreement of the promoters of the material/system. This testing must be relevant to the use of which the product will be put.

Site control

General comments

On larger projects, site control is exercised by the engineer and/or the architect, often with the assistance of full-time site staff. This 'site control' is normally described as 'inspection', not 'supervision'. The contractor is required to provide adequate supervision to ensure that the requirements of the contract are complied with.

It is now common practice for contractors to carry out the work through sub-contractors by means of separate sub-contracts. The main contractor is responsible for the supervision of the work of the sub-contractors but, unfortunately, it is not unusual to find that this responsibility has not been taken particularly seriously. Theoretically, defects have to be put right, but the result of rectified defective work is seldom as satisfactory as work that was carried out properly the first time. The legal responsibility of main contractors for the quality of the work and/or materials of nominated sub-contractors and suppliers is not open to simple interpretation and in the end may have to be referred to solicitors and counsel. In certain cases, the engineer/architect may not have the necessary experience to design and control a specialist activity that forms part of the project. In these circumstances the professional man should make this quite clear, in writing, to the client, and make sure that his fees do not include work for which he is passing the responsibility to others. There are cases where an engineer/architect is only briefed to produce the design, drawings and contract documents and is excluded from any site control responsibility. Such a situation needs to be very carefully described in writing and in any large project legal advice should be obtained on the drafting.

One way to help reduce potential allegations of defective work when the project is complete is to carry out selected non-destructive testing *during* construction.

Quality of materials

Where there is a British or other National Standard for a material such Standards include requirements for testing and, provided the importance of the project justifies the expense, it is advisable to carry out some testing in accordance with requirements in the Standard *before* the materials are incorporated into the project. For example, the grading of aggregates when concrete or mortar is mixed on site can only be effectively checked before they are batched with cement and water, because testing is directly associated with sampling. For ready-mixed materials, such sampling should be carried out at the mixing plant.

Again, on large projects, factory-made materials, such as ceramic and terrazzo tiles, concrete paving slabs and blocks, ceramic pavers, should all be sampled and tested in accordance with the relevant Standard, before being incorporated into the works. Allegations are frequently made that reinforcement has been incorrectly fixed, resulting in inadequate cover. The cover should be checked before the concrete is placed and then again as soon after placing as is practical – not 3 years later.

The ICE (Institution of Civil Engineers) Conditions of Contract require that daywork sheets be signed by the engineer's representative each day; there is no such provision in the JCT (Joint Contracts Tribunal) forms. Without the closest control, permission to use daywork is like giving a blank cheque to the contractor.

Quality of workmanship

There is now a British Standard for workmanship on building sites, BS 8000; this was in 15 parts at the time this book was written. Parts 2, 4, 9 and 11 are relevant to the construction and finishing of concrete floors and hardstandings.

The assessment of quality of workmanship and finishes is very subjective. The author's experience is that the two fundamental factors are the clarity or otherwise of the specification and 'fitness for purpose' of the resulting construction.

Investigations and testing

Introduction

The main reason for the requirement to carry out an investigation is generally visible defects which the engineer/architect and/or the user find unacceptable, and are claimed to render the floor or hardstanding 'not fit for the purpose' and/or in breach of the 'conditions of contract'.

The allegations may arise during the period of the contract, including the defects liability period, or some time later but within the period of limitation (6 years for ordinary contracts and 12 years for contracts under seal, for breaches of contract). Litigation may well be instituted as a result of the subsequent investigation. Great care is therefore necessary as the other party will seek to show that the investigation and resulting diagnosis and recommended remedial work are unsound and unjustified, on both legal and technical grounds. For this type of investigation it is strongly recommended that every reasonable effort be made to obtain agreement from all parties who are involved and are likely to be involved in the dispute, to all the details of the investigation and testing.

An investigation may be required to assess the load-carrying capacity of a floor slab, and also the suitability of the floor to withstand abrasion and chemical attack, both arising from proposed change in use. In such cases litigation is unlikely.

It is not practical to enumerate all the possible defects that can occur in concrete floors, the finishes, and external concrete paving. A fundamental division is between structural defects and non-structural defects.

It is convenient to divide non-structural defects into two categories; those in the concrete base slab of floors and those in the flooring (topping, tiles, etc.). Diagnosis of defects in external paving will be dealt with separately. A short note is given below on structural defects. Most investigations include some degree of sampling and testing.

Structural defects – floors

Structural defects (i.e structural inadequacy suggesting an inability to support safely the loads imposed) seldom occur, except when a change of use is involved.

In the case of ground-supported slabs, defects usually arise in the sub-base and/or the sub-grade, and generally show themselves as undulations in the bays, i.e. significant differences in levels between adjacent bays. In extreme cases, individual slabs may 'break their backs'. The investigation will then concentrate on the sub-base and sub-grade, but should not be confined to these two items; the concrete slab should also be investigated for quality and bay layout.

Certain types of cracks are considered as 'structural' because if they are not remedied they may result in the break-up of the slab and require more than simple sealing/grouting. Structural defects in suspended slabs may arise from the design and/or construction of the slab itself or from some inadequacy in the supporting structure. Such defects show as excessive deflexion of the slab and tension cracks in the soffit and in the top of the slab over supports.

The investigations should include checking the design if this is available; load testing may be required to decide finally whether or not strengthening is required.

Prior to load testing, it is usual to carry out one or more of the following types of testing:

- non-destructive testing using the rebound hammer, ultra-sonic pulse velocity (UPV), cover meter;
- coring, followed by laboratory examination, to determine compressive strength, mix proportions, the presence of sulphate attack, and the concentration of chlorides.

Some brief comments are given at the end of this chapter on certain aspects of laboratory testing.

Non-structural defects – floors

The base slab – directly finished
These generally consist of:

    (a) dusting of the surface and excessive wear/abrasion,
    (b) opening of joints and damage to joint arrises,
    (c) shrinkage and thermal contraction cracks,
    (d) crazing,
    (e) out of tolerance on surface levels,
    (f) slipperiness-low frictional resistance.

Preliminary investigation *of surface dusting and excessive wear* should include rebound hammer tests on each bay. A minimum of 15 readings should be taken; the highest and lowest readings being discarded if they are significantly outside the remainder. Details of the concrete mix and the method of laying, compacting, finishing and curing should be obtained if at all possible, as well as the contract specification if there is one. Coring may have to be resorted to in order to determine the compressive strength and standard of compaction, and to provide samples for chemical analysis to determine the mix proportions. An accelerated wear test using equipment developed by the British Cement Association can give useful information on the probable wearing properties of the surface of the concrete floor (see Figure 2.12). Some authoritative publications are listed in the Bibliography at the end of this chapter.

With regard to the *opening of joints, damage to joint arrises*, and *shrinkage and thermal contraction cracks*, the initial investigation should be able to determine whether further expensive investigation and probably testing is justified. This will depend on the extent to which the joints have opened over the period between the completion of the floor and the start of the investigation. The principle also applies to cracking. Crack widths and the crack pattern are important in helping to arrive at a practical assessment of the problem. Information on the original design of the floor may be required.

It is the author's experience that *crazing* is not a defect that is likely to affect the long-term durability of the floor unless it is accompanied by depressions in the surface. There has been considerable research and many papers have been written on the surface crazing of cement-based materials. If the proposed use of the floor makes crazing unacceptable (extreme hygienic requirements), this should be made clear in the contract specification. It is not at all unusual to find crazing on the surface of high-strength concrete floors.

When the surface of the base slab is alleged to be *out of tolerance* on surface level, the matter can be a very serious one, as rectification is difficult and expensive. The complaint can be that the gradient is inadequate, resulting in ponding; or is too steep, causing slipping. Another complaint can be that the surface is not sufficiently flat to allow the safe and efficient operation of high-lift turret trucks. Detailed scrutiny of the contract requirements relating to floor surface tolerances is the first step. Depending on the result of this perusal, it may be necessary to carry out a detailed survey of the floor surface. This is best done with special equipment which records on paper the profile of the floor; an experienced operator is essential.

Reference has been made in Chapter 2 to the problem of slippery floors which can be serious, especially in industrial buildings. This usually has nothing to do with the wearing properties of the floor, but an investigation is sometimes called for. There is a conflict between the need for a floor surface which is easily cleaned and a surface which is slip-resistant. An investigation should include a careful study of the work carried out on the floor and the type of footwear used by the work-force. Readers are referred to the section in Chapter 2 where this subject is discussed in some detail. The use of special equipment for measuring the frictional resistance of the floor surface is usually advisable. The 'TORTUS' developed by British Ceramic is shown in Figure 2.13

The base slab – with bonded topping

The comments given in the previous section apply here. In addition, the bond between the topping and the base concrete may require investigation. Initially, this would require checking for 'curling' at the perimeter of bays and by random tapping with a light hammer or rod to detect debonded areas by the emission of a 'hollow' sound. Limited debonding does not necessarily mean that the topping will fail in use. A decision would have to take into account the location of the debonded areas, their extent and the use to which the floor is put. A floor subjected to heavy moving loads and severe impact is much more vulnerable than one where the loading and impact is light. A sensible assessment requires considerable experience. A grid survey at 1.00 m centres in both directions is generally desirable.

Screeds and finishes

Cement/sand screeds are often found to be defective and this gives rise to many disputes, some of which reach the High Court.

It should be remembered that cement/sand screeds are not suitable as a wearing surface, even for light foot traffic. The usual defects that occur in screeds are:

(a) an inadequate resistance to impact,

(b) shrinkage cracks,

(c) debonding from the substrate and curling at the edges of bays,

(d) a surface finish that is too uneven to receive thin sheet or tile flooring.

(a) The most suitable method of checking a complaint of inadequate resistance to impact is to carry out a survey with the Building Research Establishment Screed Tester. This essentially involves measuring the depth of indentation in the screed caused by the fall of a 1 kg weight from a height of 1.00 m. Details of the use of this equipment and interpretation of results are given in a number of publications, some of which are listed in the Bibliography at the end of this chapter. Depending on the circumstances of the case, the author favours a survey at 1.00 m centres in both directions over the whole floor. Experience and common sense is needed in the interpretation of the results, and reference should be made to the specification requirements (if there are any).

(b) Shrinkage cracks on their own, provided they are not present in excessive numbers and do not exceed about 0.50 mm in width, are unlikely to have any long-term effect on the durability of the screed. Also, they can be readily grouted in.

(c) Debonding, and curling of the screed at the perimeter of bays, are usually accompanied by shrinkage cracks. The combination can cast serious doubts on the durability of the screed. Tapping with a light hammer or rod will detect loss of bond (hollowness) and the use of a straight-edge across bay joints will reveal curling. In important cases, a grid survey should be carried out, at about 1.00 m centres in both directions.

It should be noted that debonding denoted by the presence of hollow-sounding areas does not necessarily indicate that the screed will fail under impact.

(d) The contract documents should be reviewed to ascertain what standard of finish the contractor was required to provide. The search may prove to be negative and reliance is then placed by the complainant on 'fitness for purpose'. However, the standard of finish acceptable to receive rigid tiles such as terrazzo and ceramic, is unlikely to be sufficiently smooth on which to lay thin sheet and tiles such as PVC.

Floor finishes: rigid tiles, and flexible sheeting and tiles

Floors finished with rigid tiles, such as terrazzo, ceramic, conglomerate marble and marble, and flexible sheeting and tiles can develop specific defects and these are discussed briefly below.

Rigid tiles (terrazzo, ceramic and marble)

It is the author's experience that defects seldom originate in the manufacture of the tiles. Discolouration/darkening of terrazzo tiles arises from excess moisture in the tiles caused by inadequate curing, or carelessness in the storage of the tiles on site (see Clause 13 and Appendix C of BS 4131: Terrazzo tiles). The main defects found in terrazzo-tiled floors are the opening of joints between the tiles, and tilting of the tiles due to the crushing of a sub-standard bed or screed below. Terrazzo tiles are comparatively thick, e.g.25 mm and over. Ceramic tiles should be selected for the loading they will have to sustain.

Investigation should consider the location of movement joints with particular reference to the supporting structure and elements that restrain movement such as columns and perimeter kerbs. The use of a separating membrane should be looked for as such a layer in a heavily-loaded floor can result in rocking of the tiles and tile bed unless the concrete slab has been specially finished with a smooth and even surface. Polythene sheets laid on the base slab without adhesive can cause serious trouble.

Joint widths should be considered, as joints should be as narrow as practicable, taking into account the type of tiles used. The method and frequency of cleaning the floor needs investigation as well as the cleaning materials used.

A characteristic of cement-based tiles, such as terrazzo is that they are subject to drying shrinkage; this, combined with shrinkage of the grout in the joints and the method of cleaning, can result in a loss of grout, which can disfigure the floor.

Conglomerate marble tiles (reconstituted marble) suffer from pitting of the surface due to the loss of small pieces of marble. This is in addition to the hazards referred to above for terrazzo and ceramic tiles, with the exception of possible shrinkage of cement-based tiles.

In all types of rigid tiles, the problem of providing a level, even surface virtually free from 'lipping' is often encountered. Tolerances are given in the relevant Standards; these should be checked with the requirements in the contract (if there are any). If there are no contract requirements, then it is a question of whether the floor is 'fit for the purpose'. As far as possible, all measurements should be carried out as recommended in the relevant Standard, and when dealing with variations in millimetres, common sense must be used. When working to such small dimensions, it is unlikely that two independent observers will obtain exactly the same results.

Flexible sheeting and tiles
The common faults include sub-standard welded joints, loss of adhesion with the substrate due to the presence of excess moisture or use of the wrong type of adhesive, depressions in the surface due to crushing of the screed below, and staining due to the use of an unsuitable adhesive, or unsuitable cleaning materials. The use to which the floor

is put should also be considered; black rubber in footwear can mark the surface. Supports for heavy furniture, the legs of chairs and the wheels of trolleys can cause indentations in these flexible materials.

External concrete paving

The faults likely to require investigation are in many instances basically similar to those in directly-finished concrete ground floors for industrial use, which have already been discussed in this chapter.

However, external paving is laid and operates under more exposed conditions than slabs inside a building and the surface finish is likely to be quite different. A brush finish to improve skid resistance is quite usual and tolerances on surface levels will be less stringent than for most industrial floors.

Disputes over the execution of the brush finish are not unusual and use of the Sand Patch method as described in the DoT Specification for Highway Works, 7th edition (clauses 1026 and 1031) can help resolve arguments, provided of course provision has been made for this in the contract specification. Another source of dispute is the location of the reinforcement, i.e. the depth below the surface of the slab. A cover meter survey is then essential, but the results must take into account construction tolerances and the testing tolerances. If coring is carried out this may reveal honeycombing in the concrete and/or voids in the sub-base. Coring is expensive and destructive and so additional means should be sought to obtain information on the extent of the honeycombing and voids.

A method that is being increasingly used to establish layer thicknesses, detect voids and locate incipient cracks, is impulse radar. The information that follows has been obtained by courtesy of G.B. Geotechnics (Cambridge, UK). The great advantage of this technique is that it is for all practical purposes non-destructive, although a limited number of cores should be taken to calibrate the equipment and verify the results. It provides a continuous picture of sub-surface conditions and thus a thorough check on specification compliance. Pulses of electro-magnetic energy (radar) are beamed into the pavement, and the results are plotted automatically. Readings are taken at about 1 m centres.

Figure 6.1 shows impulse radar equipment being used on a public highway.

Some notes on sampling and laboratory testing

The majority of investigations include taking samples and testing in a laboratory.

Fig. 6.1 Equipment using impulse radar to obtain sub-surface information on pavement construction. Courtesy: G.B. Geotechnics.

Sampling

The number of samples and their location requires careful considera-tion because the results are intended to be representative of the area of floor or hardstanding being investigated.

The author wishes to emphasize again the importance of obtaining agreement of all the parties concerned to the sampling and testing; this principle is set out very clearly in relevant British Standards, such as BS 1881 and BS 6089. The principal Standards for sampling and test-ing are:

for screeds: BS 4551
for concrete: BS 5328, BS 6089, BS 812 and BS 1881

While BS 4551 indicates clearly the number of samples and their weight that can be considered as providing reasonably representative results, the recommendations contained in BS 5328 and BS 6089 are open to various interpretations. However, useful recommendations are made in Concrete Society Technical Report 32: *The Analysis of Hardened Concrete.*

Coring should only be used as a last resort, especially in suspended slabs. The author considers that for slab on grade, two cores per bay, and for suspended slabs, one core per bay, can be considered reason-able. Each case should be taken on its merits. Sampling of screeds and slabs is a time consuming and expensive business, and in an occupied building it is very disruptive, which is a further reason for trying to obtain agreement between the various parties prior to commencing the investigation.

Laboratory testing

Introduction
It is obviously important for the engineer in charge of the investigation to select a laboratory with experience in the type of testing required, and he should provide a clear brief to the laboratory of what is required. It is the responsibility of the engineer to interpret the test results so that he can provide a clear and sensible diagnosis and report.

Test results are not absolute as there is always a tolerance on the figures given, attributable to the precision of the test and the representativeness of the sampling. The precision of the test itself should be defined and the tolerances on the various test results should be stated in the laboratory report. Any additional sampling tolerance should be assessed separately.

Cement content
For large volumes of concrete, 10 or more samples could be necessary, as recommended in BS 1881: Part 124. The same principle applies to sampling a large number of units. If small volumes of concrete are examined (i.e. 6 m^3 or less), it is recommended that not less than four well-spaced independent samples be taken. An error of ± 25 kg of cement per cubic metre of concrete should be allowed for variability of the concrete in the structure.

In chemical analysis, to establish the cement content and thus the mix proportions of concrete and mortar, there is a considerable reduction in the sample size. Only about 5 g of powdered concrete/mortar is analysed from a combined sample of about 7 kg. The preparation of the final sample for analysis is of crucial importance to ensure that the sample analysed (5 g), is truly representative of the original concrete. There is a testing variability (tolerance) on samples tested within one laboratory and between laboratories.

Concrete Society Technical Report 32 suggests sampling variability as ± 25 kg/m^3, and testing variability as ± 30 kg/m^3, making a combined variability of ± 40 kg/m^3.

Water/cement ratio
It is sometimes necessary to determine the original water/cement ratio of the concrete. The author considers that this test is indicative only and should not be used as a 'pass or fail' test. The accuracy (bias) of the test is claimed in the Concrete Society Technical Report 32 as 0.1; this means that a result of 0.5 could mean an actual w/c ratio of either 0.4 or 0.6. This figure of ± 0.1 must be further qualified by saying that this only applies under favourable conditions where the sample is

undamaged and uncarbonated and the aggregates are known. In less favourable conditions the bias cannot be assessed accurately.

Aggregate grading

Demands are sometimes made to determine the aggregate grading from samples of concrete and mortar. Again, the author feels such tests should be considered as indicative only. The Concrete Society Technical Report 32 states clearly:

> 'It is not recommended that analysis of hardened concrete be used to establish compliance of the aggregates with the particle size requirements of the aggregate specifications.'

The correct procedure to check aggregate gradings is to take samples from the stock piles in accordance with BS 812.

Cutting and testing cores

The relevant British Standards are BS 1881: Part 120: Method of determination of compressive strength of concrete cores, and BS 6089: Guide to the assessment of concrete strength in existing structures. There is also the excellent Concrete Society Technical Report 11: Concrete core testing for strength.

For buildings under construction, core cutting and testing is usually only ordered when the cube results are deemed to be unsatisfactory after thorough investigation of sampling, and cube making and testing. When investigating an existing building it is the only way to obtain *direct factual* information on the quality of the concrete when this is required for the purpose of the investigation.

Care, commonsense and experience is needed in the interpretation of the compressive strength results; detailed study of the publications mentioned above, particularly BS 6089 and Concrete Society Technical Report 11 is strongly recommended. Core strengths and in-situ cube strengths estimated from them should be expected to be lower than standard cube strengths. This is due to the significantly different treatment (compaction and curing, and the effects of the presence of reinforcement) of concrete placed in the structure compared with the closely-controlled laboratory type procedures of preparation, compacting and curing of the standard test cubes. This is set out in detail in BS 6089 and Concrete Society Technical Report 11.

In addition to compressive strength tests, the cores can give useful information by means of a careful visual examination of both the cores themselves and the sides of the core holes. The concrete from the crushed cores can be used to check for the presence of chlorides and sulphates, and of course for the cement content, and from that the mix proportions.

A fundamental problem on which there is little guidance is the number of cores required when a large floor area has to be investigated, and there are no cube results from which basic information can be obtained. The larger the number of cores, the more reliable is the assessment that is made from the test results. Coring is expensive and destructive and so should be kept to a practical minimum.

The author suggests the following:

(a) The proposed coring should be based on recommendations for sampling in Table 15, Part 1 of BS 5328: 1991. For floor slabs this would be one core for each 20 m^3 of concrete in the slab. For a 200 mm thick slab this would result in one core for each 100 m^2, but there should be at least two cores, for an initial investigation (however, see comment in (c) below).

(b) Additional cores should be taken from any area of slab from which a core gives a noticeably low result or the concrete shows some unusual properties.

(c) When the cores are taken due to suspect cube results, the coring should be concentrated in the area of slab where the suspect batch or batches of concrete were placed. A minimum of two cores should be taken initially where the concrete under consideration is identifiable to one batch. The average result of the two cores could then be assessed to provide a test precision tolerance of ± 8.5%. Reference can also be made to clause 8.2 in BS 5328: Part 1: 1991.

Bibliography

BS 1881 Methods of testing concrete, British Standards Institution, Milton Keynes

BS 6089 Guide to assessment of concrete strength in existing structures, British Standards Institution, Milton Keynes

BS 4551 Methods of testing mortars, screeds and plasters, British Standards Institution, Milton Keynes

Building Research Establishment (1989) *Bremortest: A Rapid Method of Testing Fresh Mortars for Cement Content.* IP8/89

Chaplin, R.G. (1986) *Mixing and Testing Cement: Sand Floor Screeds,* Cement & Concrete Association Technical Report 566, p. 28

Concrete Society (1976) *Concrete Core Testing for Strength,* Technical Report 11, p. 44

Concrete Society (1988) *Concrete Core Strength Testing,* Digest No. 9, p. 10

Concrete Society (1989) *The Analysis of Hardened Concrete,* Technical Report 32

Kettle, R.J. and Sadegzadeh, M. (1986) Recent research developments in abrasion resistance. *Concrete*, Nov. 1986, 29-31

Liu, T.C. (1981) Abrasion resistance of concrete, *ACI Journal*, Title 78-29, Sept/Oct. 1981, 341–350

Pye, P.W. (1984) BRE screed tester, classification of screeds, sampling and acceptance limits, British Research Establishment IP11/84, p.4

Pye, P.W. and Warlow, W.J. (1978) *A Method of Assessing the Soundness of Some Dense Floor Screeds*, British Research Establishment, Current Paper CP72/78, p. 9

Chapter 7
Repairs, resurfacing and maintenance of floors and external paving

Introduction

In the previous chapters the author has endeavoured to record his experience in dealing with the construction and finishing of floors and external paving. The present chapter deals with repairs that are needed sooner or later. The author does not know of any type of flooring nor external paving that is likely to be free of repair for the whole of its design life, unless the design life is a particularly short one. In the context of this chapter, 'maintenance' is intended to mean little more than cleaning and small repairs that do not cause significant disturbance to the normal use of the floor.

The subject of repairs to floors and floor finishes is closely associated with the question of 'effective life' or 'useful life' of the floor or floor finish. This is a subject that gives rise to lengthy and heated discussions, sometimes resulting in litigation.

At the time of writing this book a British Standard had just been published – BS 7543; it is a guide to the practical useful life of building materials and components. This is an important document and should be closely studied. The whole concept of the effective/useful life of floors and floor finishes is very subjective indeed. The factors involved are complex and are incapable of clear definition. The main factors that determine the life span are:

(a) the extent to which environmental conditions and usage have adversely affected the ability of the floor/finish to perform its normal duties;

(b) the effect of age and use on aesthetic appearance; this is particularly subjective and depends on a number of factors;

(c) the willingness of the responsible party (tenant or building owner) to spend the necessary money for remedial work which may involve renewal. The source of the money, public funds or private purse is also an important factor; when it is believed that the cost of say renewal, can be recovered from another party, there is much more enthusiasm for insisting that complete replacement is the only satisfactory solution;

(d) the end of the 'useful life' may be determined when the cost of repairs including disruption, becomes uneconomic. Again, the term 'uneconomic' is very difficult to define clearly and unambiguously.

For the purpose of this chapter, 'repairs' are divided into three main categories:

1. Repairs to individual defective areas including joints and cracks, where the concrete slab forms the wearing surface (concrete floors and external paving).
2. Resurfacing of large areas (concrete floors and external paving).
3. Repairs to floor finishes.

The recommendations given are intended to apply to commercial and industrial premises and not to private houses.

Category 1 repairs – patch repairs to concrete floors

The first principle of repair is that deteriorated areas do not improve by being left alone – they only get worse – so the sooner such areas are repaired, the better.

The careful patching of small defective areas of a concrete floor slab can provide an acceptable solution. Usually such repairs can be carried out at the weekend when the establishment is closed or when it can be shut down with minimum dislocation of the business, e.g. during public/annual holidays.

This type of repair should generally be carried out in the following way (see also Figure 7.1):

(a) All weak and defective concrete should be cut away and all grit and dust removed. The patches should be cut out so as to be as square as possible.
(b) The surface of the concrete around the cut out areas should be cleaned and wire brushed for a distance of at least 50 mm.
(c) The surface of the cut-out concrete should be well wetted (preferably overnight if time permits). Then 20 min. before the new concrete or mortar is laid, a coat of OPC/SBR grout should be well brushed into the surface of the cut-out concrete. The mix proportions for the grout can be 50 kg of OPC to 30 litres of SBR emulsion, or such other proportions as may be recommended by the supplier of the SBR emulsion.

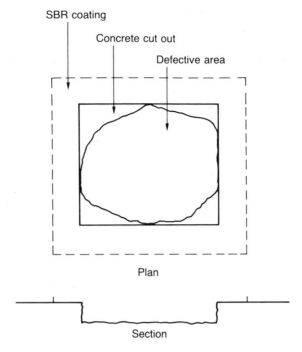

Fig. 7.1 A sketch showing a small patch repair to a concrete floor.

(d) The mix proportions for the repair concrete can be about:
 – 1 part OPC, RHPC or HAC.
 – 2 parts clean concreting sand, grade C or M to BS 882.
 – 2 parts of 10 mm coarse aggregate.
 – 10 litres SBR to 50 kg cement (about 1.25 ft^3); the same cement should be used as in the mix. Portland cement should not be mixed with high-alumina cement.
 – The amount of gauging water should be sufficient to give a slump of about 50 mm ± 25 mm.
 – If HAC is used, the w/c ratio must not exceed 0.40, and preferably should be lower.
(e) If mortar is used for the repair (for shallow patches), the mix proportions would be about:
 – 1 part cement (as for the concrete mix above).
 – 3 parts clean concreting sand (as for the concrete mix above).
 – 10 litres of SBR to 50 kg cement (about 1.25 ft^3).
 – The mix can be quite stiff and a w/c ratio of about 0.35 is likely to be adequate.
 – A bonding aid should be used as described for concrete.

For the small patch repairs considered above it is assumed that the materials will be batched by volume.

Both concrete and mortar patches require curing by covering with polythene sheeting for at least 3 days, but see below for protection from traffic.

If OPC is used the repaired areas should protected against traffic for at least 5 days; with RHPC this can be reduced to 3 days; with HAC this can be reduced to 24 h.

As long as possible after completion of the patches, a coat of cement/SBR grout should be well brushed into the dampened surface of the new repair plus the 50 mm perimeter strip of cleaned and wire-brushed concrete adjacent to it. The cement for this grout should be the same type as that used in the repair, and the mix proportions should be 30 litres of SBR to 50 kg cement. Satisfactory patch repairs can also be carried out using proprietary concrete repair kits. These can be obtained complete with bond coat material and the prepacked mortar, to which only water need be added. Special care in the preparation of the concrete and in the mixing of the prepacked materials is emphasized by the suppliers, such as the Renderoc System supplied by Fosroc, Birmingham. Renderoc S can be applied in a thickness range of 5–100 mm.

Category 1 repairs – repairs to joints in concrete floors

Floors that carry moving loads from trolleys and fork-lift trucks usually show deterioration at the joints before wear appears on the surface within the bays. This usually shows as a break-down of the arrisses. Similar deterioration usually appears along the edges of wider cracks. Repairs to cracks are dealt with in the next section. Some investigation is required before a decision can be taken on the type of repair needed as a number of factors are involved:

(a) the use to which the floor has been put and whether any change in use is contemplated (e.g. from a 'dry' to a 'wet' trade);
(b) the weight of moving loads using the floor and type of wheels, solid rubber, nylon, polyurethane and steel;
(c) the temperature range within the building;
(d) the type of joints, e.g. expansion, contraction or stress-relief.

The author's experience is that some joints are not really required once the initial contraction and drying shrinkage has taken place; when such joints start to be a maintenance problem they can be cleaned out, the edges repaired and the joint sealed with an epoxy mortar. Each case has to be taken on its merits and general rules cannot be laid down, but the 'locking' of any joints should not be considered within about 3 years of the construction of the floor.

When it is decided to maintain the joints, they can be repaired in the following manner:

(i) The ravelled and defective edges should be cut away with a concrete saw.

(ii) All sealant and back-up material should be removed and the joint carefully cleaned out.

(iii) The sides and edges of the joint should be remade with an epoxy mortar, so as to provide the predetermined joint width. The contact surfaces of the concrete should be primed as directed by the supplier of the epoxy resin prior to the application of the repair mortar.

(iv) New back-up material, debonding strip, and new sealant should then be inserted into the joint. Some information on back-up materials and sealants is given in Chapters 1, 2 and 3.

Where the floor is traversed by heavy moving loads it is worthwhile to consider protecting the joint arrisses by the insertion of stainless steel angles, as flexible sealants give very little support to the edges of the joints and damage is likely to re-occur. Depending on the use to which the floor is put, the provision of a stainless steel cover strip across the joint, fixed on one side of the joint only, and extending 75 mm each side of the joint, can be a practical solution. Careful detailing of this type of joint is essential.

Category 1 repairs – repairs to cracks in concrete floors

The presence of cracks in a concrete floor can be misleading as to their effect on long-term durability, i.e. the ability of the floor to perform the function for which it was originally intended, or for any proposed change in use. It is therefore essential to diagnose the cracks, i.e. to decide the likely cause and whether they are 'live' or 'static' and whether they are shallow surface cracks or extend through the full depth of the slab. This may involve the cutting of small diameter cores on the line of the crack(s).

There is one type of cracking which, in the experience of the author, does not require repair except in special circumstances arising from the use of the floor, i.e. 'map cracking' or crazing of the surface. This type of surface cracking is not at all unusual in concrete floors and is a type of very early drying shrinkage in the thin upper layer of the concrete. Care must be taken to ensure that map cracking due to an alkali-silica reaction is not confused with crazing.

The likely primary causes of crazing are:

(a) the use of aggregate containing excessive dust;
(b) the use of an excessively workable mix;
(c) over-trowelling;
(d) inadequate curing.

The author's experience suggests that the main cause is usually over-trowelling. The other causes ((a), (b) and (d) above) are likely to result in a rather weaker surface layer which may give rise to excessive dusting and may wear badly. Therefore the solution is essentially to determine the likely cause of the crazing and thus whether it will result in dusting and premature wear.

Crazing can and does occur in hard and abrasion-resistant surfaces and does not necessarily result in premature wear. A crazed surface may not be acceptable for a floor in a food-processing establishment, nor in the pharmaceutical industry, due to possible harbouring of micro-organisms.

The cracks are very fine, usually in the range of 0.1 to 0.3 mm in width and the depth seldom exceeds 2 mm.

A suitable treatment is to clean the surface and apply a silico-fluoride hardener, such as 'Lithurin'.

The main reasons for the repair of cracks are:

(a) To prevent fretting and ravelling along the sides of the crack. This is unlikely to occur with cracks that do not exceed about 0.75 mm to 1.0 mm in width.
(b) To prevent the penetration of water, and liquids that may be aggressive, into the concrete, or which otherwise may be unacceptable from a hygiene point of view, or which may cause dampness or damage in rooms/areas below the floor.
(c) To seal completely and permanently the cracks prior to the laying of a thin-bonded topping. This can only be successful if the cracks are not 'live' and further movement across the cracks is not anticipated and does not occur, otherwise the cracks will be reflected through the new topping.
(d) Cracks in a thin-bonded topping which are accompanied by loss of bond adjacent to the cracks require special consideration and treatment, and are discussed later in this chapter.

If it is decided that cracks in the floor slab are 'live,' in other words, movement across the cracks will take place after repair, then the repair material must be sufficiently flexible to accommodate this movement and must bond to the concrete. A guarantee of success for this type of crack repair will be difficult to obtain.

For cracks up to about 1.5 mm wide, there is generally no need to cut out the crack. The recommended procedure is to tap along the edges of the crack with a chisel to detect any weak edges, then thoroughly clean out the crack with compressed air and inject into the crack a low-viscosity polymer resin. For the smaller jobs, the resin injection can be replaced with a cement/SBR grout.

Wider cracks should be cut out, and then carefully cleaned as described above. If a cement-based mortar is to be used for filling the crack, then the crack should be wetted overnight and all surplus water blown out by compressed air before the repair mortar is inserted. A mix of 1 part cement to 2 parts of clean fine sand, with 10 litres of SBR emulsion to 50 kg of cement should be suitable. The repaired crack should be covered with polythene sheeting for 3 days or a sprayed-on curing membrane applied. The use of HAC instead of Portland cement will enable traffic to run over the repaired crack after 24 h (see Figure 7.2).

Fig. 7.2 Equipment for cutting out a wide crack in in-situ concrete paving prior to sealing. Courtesy: Errut Ltd.

Category 2 repairs – resurfacing of concrete floors

There are many reasons why a deteriorated concrete floor slab can best be dealt with by the provision of a completely new surface. The main reasons include the following:

(a) widespread deterioration;
(b) due to a change in use, a different type of surface is required, which may have to incorporate drainage and improved surface gradients, be watertight and/or be chemically resistant, or be anti-static;
(c) changes in the levels of adjacent floor areas, which require the overall level to be raised.

The provision of a completely new surface to a floor, especially to a suspended slab, must take into account the resulting increase in dead load; this can be a difficult problem to overcome. The use of artificial lightweight aggregate concrete may help to solve the problem, but such concrete has a lower resistance to abrasion than dense aggregate concrete. Where there is widespread deterioration of the existing floor surface, overslabbing is generally the best solution; detailed recommendations are given in Chapter 3. If the size of the job justifies the extra cost of strict full-time site control, consideration can be given to what is known as 'Fast Track' concrete overlay, which is new to the UK but is well known in the USA. This method is briefly described in the next section.

As the minimum thickness of a concrete overslab is likely to be 100 mm, and may well be more, the effect on the finished floor level in relation the levels of adjacent floors must be taken into account, as well as the appreciable additional dead load.

Depending on the nature and extent of the deterioration, the provision of a bituminous-based topping may also be suitable, and some details are given in Chapter 3.

Thin-bonded cement-based toppings rely for their durability on the satisfactory preparation and the quality of the base concrete, as well as on the quality of the completed topping. General recommendations for thin-bonded toppings are given in Chapter 3, to which reference should be made. A key factor in deciding on the use of a thin-bonded topping, both cement-based and resin-based, is an accurate assessment of the condition and quality of the base concrete.

The nominal thickness of such cement-based bonded toppings is likely to be in the range of 15–40 mm and so the levels of adjacent floors must also be considered. If polymer resin toppings are used, the thickness will be much less, generally in the range of 4–9 mm. These are proprietary materials, relatively very expensive, and their use is only justified in special cases. The main resins used are epoxide and polyurethane. The proprietary Ucrete toppings are laid to finished thicknesses of 6–9 mm, unless recommended otherwise by the suppliers.

Category 1 and 2 repairs to external concrete paving

While the basic principles described above for floors are generally applicable to external concrete paving, there are some important differences. External paving is subjected to a much harsher environment than floors in a building (excluding chemical attack from aggressive trade liquids). The need for water tightness seldom arises (as it may do with a suspended floor slab). Apart from wear of the concrete surface arising from the use of a too low a grade of concrete, cracking due to sub-base failure and inadequate expansion joint assemblies, the main defect is frost damage. This last defect arises from the use of a non-air-entrained concrete. If the main concrete is not air-entrained, there is little point in using air-entrained concrete for patch repairs.

The actual details of how the repair should be carried out will depend largely on the extent of the damage and the type of traffic using the paved area.

All defective concrete should be cut away, the edges of the cut-out area should be as vertical and square as possible, and all grit and dust should be removed. The mortar or fine concrete (depending on the depth and area of the repair) is usually modified by the addition of an SBR or acrylic emulsion. The recommendations given previously in this chapter for floors should be followed. For larger areas, the addition of polypropylene or stainless steel fibres, in the proportions recommended by the suppliers, will help to control cracking.

For resurfacing in concrete, either by overslabbing or by a thin-bonded topping, the use of air-entrained concrete is essential. It is important that the free water/cement ratio should not exceed 0.5 and the amount of entrained air should be in the range of about 6% ± 2%.

Bituminous and asphaltic overlays (carpets) can be very effective and are used a great deal for highways, runways and parking areas, but their useful life is much shorter than concrete.

'Fast track' construction of concrete paving

'Shut-down' time, necessitated by the need to carry out large-scale repairs/resurfacing work to external paving, access roads and hard-standings, causes disruption and financial loss. Therefore there is great pressure to reduce this shut-down period to a reasonable minimum.

With the use of 'standard' type concrete toppings, the completed topping must be protected from vehicular traffic for at least 14 days in mild/warm weather and appreciably longer in cold winter temperatures, unless special techniques are adopted. An alternative to a concrete topping is asphalt, bitumen or dense tar surfacing, with the advantage that the area can be opened to traffic within about 24 h. However, the useful life of such toppings is much shorter than that of concrete. This problem led to the development in the US of what is now known as 'Fast Track' construction in concrete. This is based on established concrete technology, and when correctly used the resulting concrete pavement can be open to traffic after 24–72 h.

Fast Track concrete mixes do not require the introduction of special materials, but material selection does require considerable care. Mix design and thorough laboratory investigation are needed to ensure that under site conditions the concrete will develop the required strength within the specified period.

The early strength of any Portland cement based concrete mix is dependent on:

(a) cement content, cement fineness and chemical composition;
(b) the water/cement ratio;
(c) the type, grading and particle shape of the aggregates;
(d) the temperature of the concrete and the ambient air temperature, and method and efficiency of curing;
(e) the degree of compaction of the concrete.

The above list shows that the basic principles employed have been known for many years. What is new is the ability to use these principles to produce very high early-strength concrete in the field.

The practical application of Fast Track concrete pavements has been successfully developed in the USA and at the time of writing this book the use of Fast Track has started in the UK. In the US it appears customary to use a Type III Portland cement (in the UK this would be a rapid-hardening Portland Cement to BS 12). The replacement of part of the cement content with PFA or GGBS is generally not recommended. In the US, an air-entraining admixture and a water-reducing admixture are used.

The use of a uniformly graded aggregate (both fine and coarse) is strongly recommended. In the UK a limestone coarse aggregate is considered advantageous as this facilitates early sawing of joints and the coefficient of thermal expansion is lower than that of other crushed rocks and flint gravel.

In the US and in the UK it has been found that a cement content of about 400 kg/m^3 and a free water/cement ratio of 0.40 should be suitable.

Literature from the US emphasizes the need for careful curing and thermal insulation. The application of a curing compound at 1.5 times the standard rate is recommended.

The sawing of joints should take place while the temperature of the concrete slab is still rising as this helps to prevent premature cracking.

The sealing compound must be suitable for use when the concrete is still damp.

All the above must be taken into account in the preliminary testing and investigation.

Repairs to cold-store floors

In Chapter 3 some basic information was given on the construction of the floors of commercial cold stores, some of which operate at temperatures of –25°C to –30°C. The repairs to such floors present considerable problems as it is very seldom that the store can be put out of operation and the temperature allowed to rise to near ambient.

On the assumption that the floor was originally well constructed, repairs are unlikely to be required for several years, by which time the concrete slab will be at a temperature well below 0°C, in other words it will be a condition fairly described as 'permafrost'.

If flexible joint sealants were used originally, these sealants will no longer be flexible in the frozen state. Unless the cold store is undergoing a complete renovation, it is reasonable to assume that the repairs will have to be undertaken with the store in normal operation, and at operating temperature.

A method of effecting such repairs is to carry them out under cover of what is known as an igloo, so that the temperature of the concrete surfaces against which the repair material is to be placed can be raised to the extent needed to suit the repair material being used.

One such material which the author understands has proved successful for this work in Certite Winter Grade, made by SBD Ltd, Bedford. Certite is a polyester resin compound and directions for use should be obtained from the suppliers.

Another material, made by Colebrand Ltd, London, and which is claimed will cure at –25°C, is based on a solvent-free urethane; directions for its use must be obtained from the suppliers and must be closely followed.

Repairs to floor finishes

There are a vast number of finishes now available and in use for concrete floors, most of which are intended to be decorative as well as durable.

Chapter 3 contains information and recommendations for the laying of screeds, various types of toppings including in-situ terrazzo, and overslabbing. Chapter 4 deals with various types of rigid tiles and flexible sheeting and tiles. This section is intended as a general guide to repairing such floor finishes.

In-situ terrazzo

The laying of in-situ terrazzo has been dealt with in Chapter 3, to which reference should be made.

In-situ terrazzo is a highly decorative floor finish and unfortunately defects are correspondingly noticeable. Before detailed consideration is given to remedial work, it is essential that the cause of the defects should be established as far as this is possible.

The main defect is cracking and if the cracks are narrow (not exceeding 1 mm in width) it is probably better to leave them alone rather than attempt repair; but this is a matter of opinion and is thus very subjective. Even the best execution of the repair will be clearly visible, as it will be impossible to match the adjacent terrazzo exactly.

Fine cracks usually occur a considerable time before they are reported, because they are not noticeable until they become discoloured with dirt.

A special type of cracking is surface crazing, also known as 'map cracking', which is a network of very fine cracks, usually less than 0.3 mm wide and not exceeding about 1.0 mm deep. The author does not know of a certain remedy, but grinding and polishing, followed by the application of a colourless sealant may effect an improvement in the appearance of the floor. The execution of a small trial area is recommended. Crazing is unlikely to affect the wearing properties of the floor, only its appearance. However, for reasons of hygiene, crazing can be unacceptable in hospitals, and in the food and pharmaceutical industries.

Repairs should only be entrusted to an experienced firm, and recommendations for suitable firms can be obtained from the National Federation of Terrazzo, Marble and Mosaic Specialists, at PO Box 50, Banstead, Surrey SM7 2RB, UK.

When cracks are numerous and wide, particularly when accompanied by debonding and defects in the control joints, the most satisfactory solution is likely to be complete renewal of the floor surface. Patching will be very unsightly.

Tiles – terrazzo

The laying of terrazzo tiles is dealt with in Chapter 4.

Before repairs of any magnitude are decided on, it is advisable to determine, as far as this is possible, the likely cause of the defects. For example, cracking and opening of joints is often the result of inadequate movement joints in the tiling. Unless this cause can be remedied, repairs will only last a limited time.

The usual defects found in use are cracking, opening of joints, and subsidence of individuals tiles due to inadequacies in the load-carrying capacity of the tile bed and/or screed, or lack of care in ensuring that the tiles are fully bedded.

The decision to repair will depend on aesthetic considerations as well as the effect of the defects on durability of the floor as a whole, and the effect on the fitness for purpose of the floor.

It must be remembered that floor tiles which have been in use for many years will have changed slightly in colour and where they have a distinctive pattern, it may not be possible to obtain new tiles with exactly the same pattern. Even with plain tiles, the new tiles will inevitably look different to the old ones. A small amount of 'lipping' between adjacent tiles can usually be remedied by grinding and polishing.

More serious unevenness of the tiles may indicate defects in the tile bed or screed, and/or workmanship in laying the tiles. Where the tile bed and/or screed is shown to be inadequate to carry effectively the loads on the floor, the tiles which are out of level must be removed together with the bedding.

The screed below the bedding may also have to be replaced. Depending on the number of defective tiles and their relative locations, an assessment may be required of the overall condition of the floor and the need or otherwise for complete replacement. This can be a very difficult decision to make, particularly in borderline cases. A relevant factor is the length of time the floor has been in use, the type of use, and the cost of replacement.

All floors need maintenance, and the defects found may be due to normal wear and tear, rather than to some latent defect. One method is to divide the floor into areas of say 2 m × 2 m and to remove carefully a tile from each area and examine the tile bed, and screed (if there is one). If the tile bed/screed is shown to be of doubtful quality, then it would not be unreasonable to recommend either complete replacement or to try resin injection which would be much quicker and far less disruptive to the normal use of the floor. It would be necessary to work out with the specialist contractor a method for checking the effectiveness of the injection. The main objection to resin injection is that holes have to be drilled through the tiles, usually at joints. There is also the problem of making good and trying to match the colour and general appearance of the replacement tiles with the existing ones. Resin injection is very expensive, but can be done quickly.

Regarding grouted joints disfigured by shrinkage cracks which have become filled with dirt, the author does not know of a proven method of remedying this unfortunate state of affairs. An attempt can be made by regrouting using a proprietary non-shrink grout, in the hope that this will mask the majority of the unsightly cracks.

Renovation of defective movement joints has been discussed earlier in this chapter, with particular reference to protecting the edges of the joints. There are a number of proprietary systems on the market, some of which are visually attractive and effective. There are diagrams illustrating these joint assemblies in the relevant Code of Practice, i.e. BS 5385, Parts 3 and 5.

Tiles – marble and marble conglomerate

The discussion above relating to terrazzo tiles is also applicable to marble and marble conglomerate tiles. However, if grinding and polishing is considered to remedy lipping between marble conglomerate tiles, special care must be exercised as small thin pieces of the marble may be removed by the grinding/polishing. Also, the grinding/polishing may so reduce the thickness of some of the small thin pieces of marble that they are likely to become loose and detached under wear and mechanical cleaning. Before any decision is taken for grinding, a small trial area should be carried out.

Tiles – ceramic

The main types of defects and principles of repair described above for terrazzo tiles are generally applicable to ceramic tiles. One exception is that ceramic tiles cannot be ground and polished to eliminate slight lipping, so that this defect must either be accepted or the uneven tiles must be taken up and re-laid. Another defect which sometimes occurs in ceramic tiles is crazing of the glazed surface. This can be the result of some irregularity in manufacture or it may be due to some inadequacy in laying, or to the use of the wrong type of cleaning materials. The crazing of ceramic tiles is a complex matter and advice should be sought from such organizations as the British Ceramic Tile Council or the British Ceramic Research Limited (both of which are based at Stoke-on-Trent, UK).

It is doubtful whether resin injection through ceramic tiles to stabilize the bed is a feasible proposition, although access via the joints would be easier due to the increased width of the joints (3–6 mm or more for ceramic tiles compared with 2–3 mm for terrazzo tiles).

Cement/sand screeds

While screeds are not suitable as a wearing surface, they form the substrate to many floor finishes, such as tiles and sheeting. Serious defects in the screed are almost certain to cause similar defects in the finish, and it is known that there are many justified allegations of failures of screeds.

The main defects are cracking, curling of the screed at the perimeter of bays, debonding (hollow-sounding areas), and crushing of the screed accompanied by the formation of depressions. When these depressions occur under vinyl sheet and tiles they are known as 'elephant foot-prints'.

Defects in screeds below rigid tiles such as terrazzo, ceramic and marble, have been discussed in previous sections of this chapter, and so this section will deal with screeds below flexible floor coverings such as vinyl and linoleum.

Before a decision is taken on the method of repair, it is most advisable to ensure that the causes of the defects have been correctly diagnosed and a practical assessment made of the likely effect of the defects found on the long-term performance of the floor under existing or anticipated conditions of use. The author's experience is that it is not unusual to find that a decision has been taken to remove completely and replace a screed when in fact, it could have been repaired satisfactorily at much less cost and much less disruption.

Therefore the types of repair that have proved successful will be briefly described, but their application to a particular set of conditions has to be left to the judgement of the professional man involved.

Apart from complete removal and renewal, there are two main categories of repair:

(a) patch repairs, or
(b) impregnation of the screed material and/or resin injection below the screed to eliminate hollowness.

Patch repairs

For patch repairs there is a choice of cement/sand polymer-modified mortar or proprietary materials of which there are a large number on the market.

Cement/sand mortars can be made with either OPC, RHPC or HAC. The sand should be a clean concreting sand to grading M in Table 5 of BS 882: 1983. The water/cement ratio should not exceed 0.40, and with the use of an SBR emulsion (10 litres to 50 kg cement), it is likely that the w/c ratio need not exceed 0.35. The mix proportions should be 1:3.5 to 1:4.0 by mass, i.e. about 1:3 or 1:3.5 by volume.

If proprietary materials are used they should be obtained from well-known firms and used in accordance with the directions of the suppliers.

The defective areas should be carefully cut out with a concrete saw; percussion tools should not be used as this is likely to damage the bond between the screed and the substrate in the surrounding sound areas. The repaired areas should be cured for 3 days and not trafficked for 5 days, except when HAC is used when the time can be reduced to 24 h.

Impregnation of the screed with low-viscosity resins

Low-viscosity resins are all proprietary systems of which there are a number of the market. They can be used with injection methods to improve bond at the interface between the base slab and the screed. The impregnation of significant areas of the screed with low-viscosity resins is intended to improve its strength and impact resistance. One of the best known systems is Flowcrete. The effectiveness of the treatment should be checked by the BRE Screed Impact Tester, and by taking a number of 50 mm diameter cores in random locations to ascertain the actual depth of resin penetration so that this can be compared with the estimated depth.

It is usual for the impregnated screed to be finished with the application of self-levelling polymer compound. This is needed because the surface of the existing screed has to be properly prepared by removal of all contamination, e.g. adhesive.

Before placing a contract for this highly specialized work, a clear method and result statement should be agreed between the employer and the contractor; this should include the nominal depth of resin penetration and the maximum indentation resulting from the use of the BRE Screed Tester. Details of the use of the tester and a realistic interpretation of results are given in the BRE publication, by P.W. Pye, IP11/84, *BRE Screed Tester: Classification of Screeds, Sampling and Acceptance Limits*. It must be emphasized that resin impregnation may cost more than removal and relaying, but there is a great saving in time and disruption.

Magnesite (magnesium oxychloride) flooring

There are three basic types of defects:

(a) cracking, arising from joints or cracks in the substrate;
(b) breakdown (softening) due to impregnation with moisture;
(c) small defective areas due to some shortcomings in the batching/ mixing of the magnesite.

When the magnesite has to be cut and replaced, it is important that this should be carried out carefully, without the use of percussion tools. Magnesite toppings are comparatively thin (16-20 mm) and rely on good bond with the substrate; this bond, although it is generally very good, may be damaged by Kango hammers and similar tools.

Cracking

Magnesite is not vulnerable to shrinkage nor thermal cracking. When cracks appear in a magnesite floor these arise from either cracks in the concrete base slab, joints in the base slab across which movement has occurred, or the failure to provide a stress-relief joint in the magnesite where the floor slab passes over supports (suspended floors).

Fortunately these cracks can be fairly easily repaired, but the work must be done by a firm experienced in the use of the material, preferably the original contractor.

Softening due to impregnation with moisture

Magnesite is vulnerable to excessive moisture, arising either from the use of the floor for a wet trade (where the floor is frequently flooded with water), or rising moisture from the sub-soil. The softening of the magnesite in the latter case is only likely where there is no damp-proof membrane in the floor system (ground-supported slabs) and the surface of the magnesite has been 'inadvertently' sealed, thus preventing the escape of moisture by evaporation from the top surface of the slab. The solution is to remove the surface seal and replace the failed areas of magnesite. A magnesite floor is a 'breathing' floor (like concrete) and so can give good service without a dpm provided the water table is not too high and the amount of moisture moving upwards is not excessive. A decision of whether a dpm is required can only be taken after a careful investigation and the results assessed in the light of experience.

Errors in batching and mixing the magnesite

The author's experience is that defects caused by batching and mixing very seldom arise but faults due to workmanship cannot be eliminated completely. The solution is to replace the defective areas.

Corrosion of embedded ferrous metal

The solution of magnesium chloride used for the gauging liquid will be aggressive to unprotected ferrous metals, and corrosion can result which would disrupt the topping. While this is a technical fact, the author must say that he has not come across a case where this defect has appeared in a magnesite floor topping.

Maintenance of floor surfaces

In this context, 'maintenance' is intended to mean regular cleaning and other processes needed to help prevent deterioration of the floor surface.

Concrete

Concrete is an absorbent material and so becomes stained very readily. While the application of surface sealants will help, care must be exercised to avoid the production of a slippery surface. The daily removal of gross contamination such as spillage of oil, grease, wax, etc., is essential. There is a wide range of industrial floor cleaners on the market, some of which are acid based and some are alkaline. The latter are generally to be preferred for use on concrete floors, as the frequent application of an acid-based cleaner can cause significant etching of the surface of the concrete.

Acid-etching or light grit blasting may be needed to restore a non-slip surface to a floor that has become dangerously smooth by heavy use. Rubber gloves and eye shields must be worn when acidic or alkaline cleaners are used. For acid etching, 1 part of commercial hydrochloric acid to 10 parts of water can be used. Reaction between the acid and the cement takes place immediately and the area should be well washed down with clear water after about 10 min. Two or more applications may be needed.

With coloured (pigmented) concrete, efflorescence can be a problem. This is also known as 'lime-bloom'. It is caused by the reaction between hydrating calcium compounds (mainly calcium hydroxide) in the cement with carbon dioxide in moist air, resulting in a very thin deposit of calcium carbonate on the surface of the concrete. The whitish stain is likely to disappear after washing but may re-appear with drying out. Unless there is a problem of rising moisture through the floor, the trouble should disappear fairly rapidly with frequent washing. This does highlight a problem that can arise with pigmented concrete, particularly the darker shades.

In-situ terrazzo and terrazzo tiles

The use of an emulsion floor wax will help in maintaining the surface in a clean condition. Cleaning with a neutral detergent or just warm water should be sufficient; soap solutions should not be used as this will make the surface slippery.

Frequent (several times daily) washing of the floor with mechanical cleaning machines incorporating rotary brushes and suction to remove the dirty wash water can cause damage to the grouted joints between terrazzo tiles.

While cases of efflorescence on in-situ terrazzo and terrazzo tiles have been reported, the author has not personally come across them.

Marble and marble conglomerate

The author is indebted to Reed Harris Ltd, London, for the following recommendations:

(a) After completion of laying, any small mortar droppings should be carefully scraped away. When the surface is free of such contamination, it should be well washed down with warm water (soap should not be used). This initial cleaning should be carried out using a mildly abrasive floor pad.

(b) A water-based polymer seal should be applied to the cleaned dry floor. New floors should receive two initial applications of the seal. Further applications of the seal should be carried out as required by the intensity of traffic using the floor.

Magnesite (magnesium oxychloride)

It is advisable for the completed floor to be sealed, but with a sealant which allows the magnesite to 'breathe'. The type of seal used should be as recommended by the specialist firm that laid the floor. The use of the wrong type of seal can have serious consequences on the performance of the floor by preventing the escape of water vapour from the top surface which can result in the accumulation of water within the magnesite leading to softening and failure.

Ceramic tiles

Reference should be made to BS 5385: Part 3: Section 7 where detailed recommendations are given for after-installation care, cleaning and maintenance.

Sheet and tile flooring (materials covered by BS 8203)

Reference should be made to the recommendations in BS 6263: Part 2: Code of Practice for the care and maintenance of floor surfaces – sheet and tile flooring. The Code offers guidance on the care and maintenance of the various materials covered by BS 8203 in particular environments and when laid in special buildings.

Bibliography

American Concrete Pavement Association (1989) *Fast Track Concrete Pavements*, Technical Bulletin TB004.0 IT, ACPA, Arlington Heights, Illinois, USA, p. 12

BS 6089 Guide to assessment of concrete strength in existing structures, British Standards Institution, Milton Keynes

BS 6213 Guide to the selection of constructional sealants, British Standards Institution, Milton Keynes

BS 8210 Guide to building maintenance, British Standards Institution, Milton Keynes

PD 6484 Commentary on corrosion at bi-metallic contacts and its alleviation, British Standards Institution, Milton Keynes

BS 6263 Code of Practice for the care & maintenance of floor surfaces, British Standards Institution, Milton Keynes

BS 7543: 1992 Guide to durability of buildings and building elements, products and components, British Standards Institution, Milton Keynes

Building Research Establishment (1979) *Estimation of Thermal and Moisture Movements and Stresses – Parts 1, 2 and 3*, Digests 227–229

Building Research Establishment (1991) *Why Do Buildings Crack?* Digest 361

Building Research Establishment 91991) *Sulphate and Acid Resistance of Concrete in the Ground*, Digest 363, p. 8

Cement and Concrete Association (1968) *Notes on the Occurrence and Repair of Faults in Granolithic Toppings*, Advisory Data Sheet No. 7, p. 5

Concrete Society (1976) *Concrete Core Testing for Strength*, Technical Report 11, p. 44

Concrete Society (1982) *Non-structural Cracks in Concrete*, Technical Report 022, p. 40

National Federation of Terrazzo Marble and Mosaic Specialists Handbook 1992, p. 40

Perkins, P.H. (1986) *Repair Protection and Waterproofing of Concrete Structures*, 2nd edn., Elsevier Applied Science Publishers, London, p. 302

Roeder, A. and Jones, D. (1991) Introducing fast track paving. *Concrete Quarterly*, Autumn 1991, 26–27

Index